MEMOIRS

OF THE

AMERICAN MATHEMATICAL SOCIETY

NUMBER 43

SPLITTING
IN TOPOLOGICAL GROUPS

BY

K. H. HOFMANN

AND

PAUL MOSTERT

Published by the
American Mathematical Society
Providence, Rhode Island
1963

First printing, 1963
Second printing with additions, 1972

International Standard Book Number 0–8218–1243–2
Library of Congress Catalog Card Number 52–42839

Splitting in Topological Groups
Karl Heinrich Hofmann and Paul S. Mostert*

Introduction

Let G be a topological group and H a closed normal subgroup. An important aspect of the study of the structure of groups is the problem of determining the degree to which G approximates the structure of $H \times G/H$. Fundamental in this study is the question of determing those subgroups of G approximating such a decomposition most closely. Only in rare instances can one hope for a splitting into the direct product, although such has been obtained, for example, in the following cases: If G is a locally compact abelian group compact modulo its component and H a vector subgroup or the maximal compact subgroup [12], [23, p. 110]; if G is a maximally almost periodic connected locally compact group and H is either a vector subgroup or the maximal compact subgroup [6], [23]; if G is a connected locally compact two ended group and H is a real subgroup or the maximal compact subgroup [7], [10]. Well known is the splitting of p-primary groups in discrete abelian torsion groups and of divisible groups in abelian groups.

If G contains a subgroup K isomorphic to G/H such that $H \cap K = \{1\}$, the identity of G, and $(h, k) \to hk$ is a homeomorphism of $H \times K$ onto G, then G is said to be a *semidirect product* of H and K, and we say that H *splits in* G. (Note that if K is also normal, then the splitting is direct.) There are abundant examples to show that a semidirect splitting need not be direct. An important instance of a semidirect splitting, proved by Iwasawa, is the case where H is a finite dimensional vector subgroup and G/H is compact [9]. An elegant proof of this was given by Cartier [4]. It is this type of splitting that is the leading motive in our endeavour, and in fact our techniques yield yet another proof of this theorem.

A semidirect splitting of a group into a normal vector subgroup H and a compact group C suggests the group of Euclidean motions which splits

*This work was supported by the National Science Foundation.

this way. In fact, let $f:R^n \to H$ be an isomorphism of the additive group of n-dimensional real vector space onto H; then there is an inner product on R^n such that every mapping $x \to f(c^{-1}f^{-1}(x)c)$, $c \in C$ is an orthogonal transformation relative to this inner product. On the Euclidean space so defined the mapping $T_{hc}:x \to f(c^{-1}f^{-1}(x)hc)$ is an isometry, i.e. a translation followed by a rotation and possibly by an inflection. The mapping $g \to T_g$ is a homomorphism of G into the group of isometries of R^n. Its kernel K is the intersection of the centralizer of H with C. Now G/K is a Lie group isomorphic to a subgroup of the isometry group of R^n. But G itself, of course, need not be a Lie group even if it is finite dimensional, and K need not split.

We shall show in an appendix that Iwasawa's theorem can be generalized in the following fashion: If G is a locally compact group, H a vector subgroup such that the quotient space G/H is compact, then there exists a normal vector subgroup N and a compact subgroup C such that G is the semidirect product of N and C.

In a semidirect product $G = KH$, the mappings $\phi:kh \to h$ and $\phi^{\perp}:kh \to h$ play an important role. Observe that the latter is an endomorphism, whereas the former satisfies the relation

$$\phi(gg') = g\phi(g')g^{-1}\phi(g).$$

A function mapping G into itself and satisfying this relation will be called a *crossed endomorphism*. (Compare this terminology with [1]. This has also been called a cocycle [4, p. 1].) The basic tool in these investigations is a technique for constructing crossed endomorphisms somewhat similar to transfer in the study of abstract groups. As in [4] and [9], the use of invariant integration is essential to our work. Our method, as Iwasawa's, uses cross sections and the invariant measure on G/H, whereas Cartier's process requires no cross section but does require an invariant integral on H, so that the latter method cannot be extended to the case of a not locally compact subgroup. In a similar context, integration was recently used by Stewart [21].

Except in Sections 2 and 4, where algebraic results dominate and which are independent of the succeeding sections, we are concerned with topological groups. Although these sections are not an integral part of our topological investigations, they are closely related in that our "basic tool" is used repeatedly in the proofs of the results therein. The notion of the crossed endomorphism leads to the introduction of the *crossed radical*.

The greater part of the paper is devoted to the splitting of topological vector groups in topological groups and applications. We shall list some of the results in a convenient, but not necessarily in their most general, form in the order in which the sections containing them occur.

SECTION I. Dual pairs of endomorphisms and crossed endomorphisms are studied, and necessary and sufficient conditions for splitting are given in terms of a dual pair.

SECTION II. The crossed radical is introduced and its relation to the Bare radical [2] investigated.

SECTION III. Invariant means are defined and produced.

SECTION IV. As an example, the following known result is proved:

Let A be an abelian normal subgroup of G with finite index n in G. Then the following condition is sufficient for A to split in G: A is uniquely divisible by n. This condition gives an extension to a classical result of Schur's (see e.g. [24, p. 132]).

SECTION V. If N is a closed normal subgroup isomorphic to a reflexive Banach space, G/N is compact and N admits a cross section, the N splits in G.

SECTION VI. If G is first countable, then the cross section in Section V exists.

SECTION VII. If N is isomorphic to a Hilbert space, then G admits a 1-1 representation onto a projective limit of Lie groups in which the image of N splits.

SECTION VIII. The results of Section VII are extended from vector subgroups to certain solvable subgroups.

SECTION IX. Extensions of locally compact connected abelian groups by compact automorphism groups are described. The well known splitting of such groups into a vector group and a compact group can be so arranged that both factors are invariant under a given compact automorphism group. Several useful corollaries are proved.

SECTION X. Generalisation of the results of Sections V-VII to the case of locally compact quotient groups are derived for central extensions of a vector group.

SECTION XI. A general principle is developed of constructing nonsplitting extensions of abelian groups by abelian groups. The examples used in this section are used to illustrate and determine the scope of the results in the following sections.

SECTION XII. The previous results are applied to locally compact groups which are compact modulo their component and contain an invariant compact neighborhood of the identity. Applications are given to maximally almost periodic groups.

SECTION XIII. The structure of two ended groups which are compact modulo their component is fully described.

SECTION XIV. The notion of unimodularity is analyzed in terms of the splitting of the maximal unimodular subgroup.

Appendix: The generalisation of Iwasawa's theorem mentioned earlier is proved.

Aside from the algebraic portion, there are two dominant themes in this study: First, the extension of a known result on splitting of locally compact vector groups beyond the confines of local compactness, and secondly the expansion of the structure theory of certain locally compact connected groups to the category of locally compact groups which are compact modulo their component.

The great wealth of literature on locally compact groups is almost exclusively concerned with the case of connected groups. It has sometimes been subconsciously believed that the description of the non-connected case is an easy, if not trivial, extension, and this has led, e.g. in the category of groups we deal with in Section XII, to incorrect statements by several authors (see Section XII). It is hoped that some of the difficulties in this connection have been overcome by the results of Sections IX and XII. We call particular attention to Corollary XII.1, which is a summary of what has been achieved in the case of maximally almost periodic groups.

In recent years, a great deal has been learned about the structure of connected locally compact groups on the one hand and of finite groups on the other. The natural synthesis of these two seems to be locally compact groups which are compact modulo their component. It is the opinion of the authors that investigations in this direction would be both interesting and non-trivial.

The following notation is standard throughout the paper:

If G and H are topological groups and f a mapping from G into H satisfying $f(gg') = f(g)f(g')$, then f is called a *representation* if it is continuous and an *isomorphism* if it is also one to one and continuously invertible (i.e. if it is a homeomorphism of G into H). The mapping f is called a *homomorphism* if it is a representation and the image $f(G)$ is isomorphic to the quotient of G modulo the kernel $f^{-1}(1)$ under the obvious mapping in the sense just defined. Thus *isomorphic* always means *algebraically* and *topologically isomorphic*.

I. Crossed Endomorphisms and Splitting of Groups.

Throughout these investigations G will consistently denote a Hausdorff topological group; it should be noted, however, that the discrete topology is admitted so that many of the results hold for abstract groups (this applies particularly to this and the following section).

We shall write inner automorphisms in the exponent, i.e. if g and h are group elements then $h^g = g^{-1}hg$, and $h^{-g} = g^{-1}h^{-1}g$.

1.1. DEFINITION. A mapping ϕ of G into G is called an *endomorphism* if it is continuous and satisfies

(i) $\phi(gh) = \phi(g)\phi(h)$ for all $g,h \in G$.

A mapping ϕ of G into G is called a *crossed endomorphism* if it is continuous and satisfies

(ii) $\phi(gh) = \phi(h)^g \, \phi(g)$ for all $g,h \in G$.

(In particular cases this has been called a cocycle [4]. Our terminology is due essentially to R. Baer [1].)

1.2. PROPOSITION. *If ϕ is an endomorphism, then $\phi^\perp = 1-\phi$ defined by*

(i) $\phi^\perp(g) = g\phi(g)^{-1}$

is a crossed endomorphism. If ϕ is a crossed endomorphism, then $\phi^\perp = -\phi+1$ defined by

(ii) $\phi^\perp(g) = \phi(g)^{-1}g$

is an endomorphism. A crossed endomorphism ϕ satisfies

(iii) $\phi(g^{-1}) = \phi(g)^{-g}$.

The following identities hold both for endomorphisms and crossed endomorphisms:

(iv) $\phi^{\perp\perp} = \phi$,

(v) $\phi^\perp\phi = \phi\phi^\perp$.

PROOF. It is clear that the continuity of ϕ implies the continuity of $\phi^\perp$ in both cases since G is a topological group.

If ϕ is an endomorphism then

$$\phi^\perp(gh) = gh\phi(gh)^{-1} = g(h\phi(h)^{-1})g^{-1} \cdot g\phi(g)^{-1} = \phi^\perp(h)^g \, \phi^\perp(g).$$

This shows that $\phi^\perp$ satisfies definition 1.1.(ii).

On the other hand, if ϕ is a crossed endomorphism then

$$\phi^\perp(gh) = \phi(gh)^{-1}gh = \phi(g)^{-1}g \cdot \phi(h)^{-1}g^{-1}gh = \phi^\perp(g)\phi^\perp(h)$$

and this proves the second statement.

We prove (iii) by observing that $1 = \phi(gg^{-1}) = g\phi(g^{-1})g^{-1}\phi(g)$ and hence $\phi(g^{-1}) = \phi(g)^{-g}$. The identities

$$\phi(g) = g(\phi(g)^{-1}g)^{-1} = (g\phi(g)^{-1})^{-1}g$$

prove (iv).

Finally if ϕ is a crossed endomorphism then

$$\phi\phi^{\perp}(g) = \phi(\phi(g)^{-1}g) = \phi(g)^{-1}\phi(g)\phi(g)\phi(\phi(g)^{-1}).$$

Clearly $\phi^{\perp}\phi(g) = \phi(\phi(g))^{-1}\phi(g)$ follows from the definition of $\phi^{\perp}$, and by (iii)

$$\phi(\phi(g)^{-1}) = \phi(g)^{-1}\phi(\phi(g))^{-1}\phi(g).$$

Combining these identities furnishes the proof of (v).

1.3.   DEFINITION. If ϕ is an endomorphism or a crossed endomorphism then $\phi^{\perp}$ is called the *dual* of ϕ. The ordered pair $(\phi,\phi^{\perp})$ respectively $(\phi^{\perp},\phi)$ of dual mappings with the crossed endomorphism in the first place is said to be an *involutive pair* of mappings.

1.4.   DEFINITION. Let ϕ be an endomorphism or a crossed endomorphism. Then

(i) $G\phi = \{ g: \ \phi(g) = 1 \}$,

(ii) $\phi(G) = \{ g: \ g = \phi(h) \text{ for some } h \in G \}$,

(iii) $\phi G = \{ g: \ \phi(g) = g \}$.

1.5. PROPOSITION. *If ϕ is an endomorphism or a crossed endomorphism, then*

(i) $G\phi = \phi^{\perp}G$,

(ii) $G\phi \subset \phi^{\perp}(G)$,

(iii) $G\phi \cap G\phi^{\perp} = \{1\}$,

(iv) $G\phi^{\perp} \cdot G\phi = \{g : \phi\phi^{\perp}(g) = 1\} = G\phi \cdot G\phi^{\perp}$,

(v) $G\phi^{\perp} \cdot G\phi / G\phi$ *is homeomorphic to* $G\phi^{\perp}$ *under the mapping* $gG\phi \rightarrow \phi(g)$.

REMARK: Statement (i) implies that the kernel $G\phi$ of a crossed endomorphism ϕ is a (closed) group and that its fixed point set ϕG is a (closed) normal subgroup.

REMARK: If ϕ is an endomorphism then the homeomorphism of (v) is in fact also an algebraic isomorphism.

PROOF. (i) follows immediately from the dual pairing of ϕ and $\phi^\perp$, as does (ii).

If $g \in G\phi \cap G\phi^\perp$ then $\phi(g) = \phi^\perp(g) = 1$ and this implies

$$1 = \phi^\perp(g) = \phi(g)^{-1}g = 1g = g,$$

assuming ϕ is the crossed endomorphism.

In order to prove (iv) let $g \in G\phi^\perp$ and $h \in G\phi$. Because of the commutativity of ϕ and $\phi^\perp$ (1.2.(v)) we may, without loss of generality, assume that $\phi^\perp$ is an endomorphism. Then

$$\phi\phi^\perp(gh) = \phi(\phi^\perp(g)\phi^\perp(h)) = \phi(1 \cdot \phi^\perp(h)) = \phi(h)$$

since h is a fixed point of $\phi^\perp$. But because $h \in G\phi$ we have $\phi\phi^\perp(gh) = 1$. Conversely, if $\phi\phi^\perp(g) = 1$, then $\phi^\perp(g) \in G\phi$ and $\phi(g) \in G\phi^\perp$, for ϕ commutes with $\phi^\perp$. By 1.2.(i) we have (again assuming that ϕ is the crossed endomorphism) $g = \phi^\perp(g)\phi(g)$ and, therefore, $g \in G\phi \cdot G\phi^\perp = G\phi^\perp \cdot G\phi$.

We now prove (v). Let π be the natural projection $G\phi^\perp \to G\phi^\perp \cdot G\phi/G\phi$ mapping g onto its coset $gG\phi$. Then π is open and continuous and its restriction $\pi|G\phi^\perp$ to $G\phi^\perp$ is a one-one map onto $G\phi^\perp \cdot G\phi/G\phi$. The inverse of $\pi|G\phi^\perp$ is equal to $\phi\pi^{-1}$ which is continuous since π is open and ϕ is continuous.

1.6. DEFINITION. We say that a closed subgroup N *splits* in G if N is normal in G and there is a closed subgroup C such that
(i) $N \cap C = \{1\}$,
(ii) $G = NC$,
(iii) The mapping $(n,c) \to nc$ is a homeomorphism of $N \times C$ onto G.
In this case G is said to be the *semidirect product* of N and C. Clearly G/N is isomorphic to C.

1.7. EXAMPLE. There is an abelian topological group G having two closed algebraically and topologically isomorphic subgroups N and C such that
(i) $N \cap C = \{1\}$,
(ii) $G = NC$,
(iii) The mapping $(n,c) \to nc$ is *not* a homeomorphism of $N \times C$ onto G.
Moreover, in the example we give, although G/N (which is isomorphic to G/C) is compact and connected, N is not even locally compact or connected, nor is G.

PROOF. Let N and C be (algebraically) an unrestricted infinite product of the circle group $\{e^{2\pi i t}:\ 0 \leq t \leq 1\}$; i.e. the elements of N and C are vectors $(x_i)_{i \in I}$ with some infinite index set I and elements x_i in the circle group. The multiplication is coordinate-wise. Let G be (algebraically) the direct product of N and C. We identify N with $N \times \{1\}$ and C with $\{1\} \times C$. For any vector $\epsilon = (\epsilon_i)_{i \in I}$ of positive numbers ϵ_i let D_ϵ be the set of all $((x_i)_{i \in I}, (y_i)_{i \in I})$ in G such that for all $i \in I$ *we have*

$$|\arg x_i - \arg y_i| < \epsilon_i.$$

The following relations are easily verified:

$$D_\epsilon D_\epsilon \subset D_{2\epsilon}, \quad D^{-1} = D.$$

This collection of subsets obviously defines a topology $\mathfrak{D}$ on G which makes multiplication and inversion continuous. G with this topology has the following properties:

(a)   The closure of $\{1\}$ is the diagonal,
(b)   The quotient topologies on G/C and G/N are trivial,
(c) $\mathfrak{D}$ induces on N and C the box topology [13, p. 107].

Let $\mathfrak{T}$ be the Tychonoff topology on G and $\mathfrak{Q}$ the l.u.b. of $\mathfrak{D}$ and $\mathfrak{T}$. Then with respect to $\mathfrak{Q}$

(a′) G is a topological group,
(b′)   The quotients G/C and G/N are algebraically and topologically isomorphic to N and C respectively with the Tychonoff topologies,
(c′) $\mathfrak{Q}$ induces on N and C the box topology.

Theorem I. *Let G be a topological group and $(\phi, \phi^\perp)$ an involutive pair of a crossed endomorphism ϕ and its dual $\phi^\perp$. Then the following eight statements are equivalent:*

(i) $\phi^2 = \phi$,
(ii) $\phi^{\perp 2} = \phi^\perp$,
(iii) $\phi\phi^\perp = 1$ (= *endomorphism mapping all of G on 1*),
(iv) $G\phi^\perp = \phi G = \phi(G)$,
(v) $G\phi = \phi^\perp G = \phi^\perp(G)$,
(vi) $G = G\phi \cdot G\phi^\perp$,
(vii) $G/G\phi^\perp$ *is isomorphic to* $G\phi$ *under the homomorphism* $gG\phi^\perp \to \phi^\perp(g)$,
(viii) $G/G\phi$ *is homeomorphic to* $G\phi^\perp$ *under the map* $gG\phi \to \phi(g)$.

It should be noted that (vi), (vii), and (viii) together with 1.5.(iii) mean that $G\phi^\perp$ splits in G and G is the semidirect product of $\phi^\perp G$ and ϕG.

PROOF. (i) implies (iv): From $\phi(\phi(g))=\phi(g)$ it follows that $\phi(g)$ is a fixed point of ϕ and belongs, therefore, to $G\phi^\perp$ which proves $\phi(G)\subset G\phi^\perp$; the rest follows from 1.5.

(iv) implies (iii): If all $\phi(g)$ are in $G\phi^\perp$, then

$$\phi(\phi^\perp(g)) = \phi^\perp(\phi(g)) = 1.$$

(iii) implies (v): If $\phi(\phi^\perp(g))=1$ for all g, then all $\phi^\perp(g)$ belong to $G\phi$. By 1.5. this proves the assertion.

(v) implies (ii): Suppose that for all $g\in G$ we have $\phi^\perp(g)\subseteq G\phi$, i.e. $\phi\phi^\perp(g)=1$. Then $\phi^\perp(\phi^\perp(g))=\phi^\perp(\phi(g)^{-1}g)=\phi^\perp(\phi(g))^{-1}\phi^\perp(g) = \phi^\perp(g)$, because $\phi^\perp(\phi(g))=\phi(\phi^\perp(g))=1$ (see 1.2.(v)).

(ii) implies (i): Let $\phi^\perp(\phi^\perp(g))=\phi^\perp(g)$ for all g and compute $\phi(\phi(g))=\phi(g)\phi^\perp(\phi(g))^{-1}$ from 1.2.(i). But similarly we have $\phi^\perp(\phi(g))=\phi^\perp(g\phi^\perp(g)^{-1})=\phi^\perp(g)\phi^\perp(g)^{-1}=1$ since $\phi^\perp=\phi^\perp$. So far we have proved that (i)–(v) are equivalent.

(iv) and (v) imply (vi): $g = \phi(g)\phi^\perp(g)$ by 1.2.(ii).

(vi) implies (vii): This follows from 1.5.(v) and the remark to that proposition.

(vii) implies (iv): Trivial.

The previous two steps carry over directly to a proof that (vi) implies (viii) implies (v).

This observation accomplishes the proof of the theorem.

1.8. PROPOSITION. *If ϕ is a crossed endomorphism of G into N and ψ is a homomorphism of $\phi(G)$ into N such that it commutes with all inner automorphisms of G then $\psi\circ\phi$ is a crossed endomorphism of G into N.*

PROOF. $\psi\cdot\phi(gh)=\psi(\phi(h)^{g^{-1}})\psi(\phi(g))=\psi(\phi(h))^{g}\,\psi(\phi(g))$.

COROLLARY I.1. *Let G be a topological group and ϕ a crossed endomorphism into a closed normal subgroup N whose restriction to N is an isomorphism onto $\phi(G)$ commuting with all inner automorphisms of G. Then N splits in G; more specifically, there is a subgroup C isomorphic to G/N such that*
(i) *$G=NC$,*
(ii) *$N\cap C=\{1\}$,*
(iii) *$(n,c)\rightarrow nc$ is a homeomorphism of $N\times C$ onto G.*

PROOF. Let ψ be the inverse of the restriction of ϕ to N. Then ψ commutes with all inner automorphisms of G. Hence, by 1.8., $\phi'=\psi\circ\phi$ is a crossed endomorphism of G onto N. In addition, for all $n \in N$ we have $\phi(n) = n$. Thus, condition (i) of Theorem I is satisfied, whence the corollary.

The preceding definitions, proposition, and Theorem I have global character. Nevertheless, they all have local equivalents which we shall need on later occasions and which we now develop parallelly to our global statements. If proofs are simply copies of what we have already done, they will be omitted.

We make the following convention: If P is a property of certain group elements g and if there is a neighborhood U of 1 in G such that all elements in U have property P, then we whall say that *P holds for all g sufficiently close to 1, or equivalently, that P holds for all sufficiently small g.*

1.1. LOC DEFINITION. A mapping ϕ of a neighborhood U of 1 in G into G is called a *local endomorphism* if it is continuous and satisfies the following condition:

(i)   If g,h and gh are in U, then

$$\phi(gh) = \phi(g)\phi(h).$$

A mapping ϕ of a neighborhood U of 1 in G into G is called a *local crossed endomorphism* if it is continuous and satisfies the following condition:

(ii)   If g,h and gh are in U, then

$$\phi(gh) = \phi(h)^g\, \phi(g).$$

1.2. LOC PROPOSITION. *If ϕ is a local endomorphism, let $\phi^\perp$ be defined by*
(i) $\phi^\perp(g) = g\phi(g)^{-1}$ *whenever ϕ is defined.*
Then $\phi^\perp$ is a local crossed endomorphism. If ϕ is a local crossed endomorphism, let $\phi^\perp$ be defined by
(ii) $\phi^\perp(g) = \phi(g)^{-1}g$ *whenever $\phi(g)$ is defined.*
Then $\phi^\perp$ is a local endomorphism. A local crossed endomorphism satisfies
(iii) $\phi(g^{-1}) = \phi(g)^{-g}$ *whenever $\phi(g^{-1})$ and $\phi(g)$ are defined.*
The following identities hold sufficiently close to 1 both for local endomorphisms and local crossed endomorphisms:
(iv) $\phi^{\perp\perp}(g) = \phi(g)$
(v) $\phi^\perp\phi(g) = \phi\phi^\perp(g).$

The proof is the same as for 1.2; we have, however, to observe that expressions like $\phi(\phi(g)^{-1}g)$ are always defined for sufficiently small g because of the continuity of the multiplication and inversion.

1.3. LOC DEFINITION. From Definition 1.3 it should be clear what we intend by the *dual* of a local (or local crossed) endomorphism, and by an *involutive pair* of local mappings of U into G.

1.4. LOC DEFINITION. Let ϕ be a local endomorphism or a local crossed endomorphism defined on a neighborhood U of 1. Then

$$U\phi = \{g : g \in U, \phi(g) = 1\},$$

$$\phi(U) = \{g : g = \phi(h) \text{ for some } h \in U\},$$

$$\phi U = \{g : g \in U, \phi(g) = g\}.$$

1.5. LOC PROPOSITION. *If ϕ is a local endomorphism or a local crossed endomorphism defined on U, then*
    (i) $U\phi = \phi^{\perp}U$,
    (ii) $U\phi \subset \phi^{\perp}(U)$,
    (iii) $U\phi \cap U\phi^{\perp} = \{1\}$,
    (iv) $(U\phi^{\perp} \cdot U\phi) \cap V = \{g : g \in V, \phi\phi^{-}(g) = 1\} = (U\phi \cdot U\phi^{\perp}) \cap V$ *for some*
neighborhood V of 1,
    (v) $U\phi^{\perp} \cdot U\phi$ *is homeomorphic to the product space* $U\phi^{\perp} \times U\phi$ *under the mapping* $g \to (\phi(g), \phi^{\perp}(g))$.

REMARK. Statement (i) implies that the kernel $U\phi$ of a local crossed endomorphism is a (closed) local group, and that its fixed point set is a (closed) normal local group; more specifically, if $g, h \in U$ (resp. $g, h \in U\phi^{\perp}$), $gh \in U, g^{-1} \in U$ and $g^{k} \in U$ for some $k \in G$, then $gh, g^{-1} \in U\phi$ (resp. $gh, g^{-1}g^{k} \in U\phi^{\perp}$), and $U\phi$ (resp. $U\phi^{\perp}$) is closed in U.

The proofs of (i) through (iv) are straightforward modifications of the proofs to 1.5. In order to prove (v) it is sufficient to observe that the continuous mappings $g \to (\phi(g), \phi^{\perp}(g))$ of $U\phi^{\perp} \cdot U\phi$ to $U\phi^{\perp} \times U\phi$ and $(g, h) \to gh$ of $U\phi^{\perp} \times U\phi$ to $U\phi^{\perp} \cdot U\phi$ are inverses of each other.

THEOREM I LOC. Let G be a topological group and $(\phi, \phi^{\perp})$ an involutive pair of a local crossed endomorphism ϕ and its dual local endomorphism. Then the following six statements are equivalent:
    (i) $\phi(\phi(g)) = \phi(g)$ for sufficiently small g,
    (ii) $\phi^{\perp}(\phi^{\perp}(g)) = \phi^{\perp}(g)$ for sufficiently small g,
    (iii) $\phi\phi^{\perp}(g) = 1$ for sufficiently small g,
    (iv) $U\phi^{\perp} \cap V = \phi U \cap V = \phi(U) \cap V$ for some neighborhood V of 1,
    (v) $U\phi \cap V = \phi^{\perp}U \cap V = \phi^{\perp}(U) \cap V$ for some neighborhood V of 1,

(vi) $U\phi^\perp \cdot U\phi$ is a neighborhood of 1 and the product is topologically direct.

The proof that (i) through (vi) are equivalent again uses no new ideas beyond the corresponding proofs for Theorem I.

II.   The Radical and the Crossed Radical.

In this section, a dual pair $(\phi,\phi^\perp)$ will always satisfy the condition that $\phi(G)$ is in some abelian subgroup A of G. No topology is involved in these considerations. The results of this section are applied only in section 4.

2.1. **Proposition.** *The following properties hold for dual pairs* $(\phi,\phi^\perp)$:
(i) $\phi(ag) = \phi(a)\phi(g)$ *if* $a \in A$,
(ii) $\phi(a^g) = \phi(a)^g$ *if* $a \in A$,

(iii) $\phi^{\perp n \perp}(g) = \prod_{k=1}^{n} \phi^k(g)^{(-1)^{k-1}(n;k)}$ *where* $(n;k)$ *denotes the binomial*

coefficient $n!/k!(n-k)!$.

Proof. (i) Since A is abelian, $\phi(g)^{a^{-1}} = \phi(g)$ and the result follows from 1.1.

(ii) follows similarly with the observation that $\phi(g)^g = \phi(g^{-1})^{-1}$ (see 1.2.(iii)).

(iii) is proved by induction on n using the defining properties of $\phi^\perp$.

2.2. **Definition.** The *radical* $R\phi$ of the pair $(\phi,\phi^\perp)$ is the radical of $\phi^\perp$ [2]. That is

$$R\phi = \bigcup_{n=1}^{\infty} G\phi^{\perp n}.$$

The *crossed radical* $C\phi$ of the pair $(\phi,\phi^\perp)$ is the set

$$C\phi = \bigcup_{n,m=1}^{\infty} G\phi^{\perp n \perp m}.$$

2.3. **Proposition.** *The radical and crossed radical are groups,* $R\phi$ *is normal, and* $R\phi \cap C\phi = 1$. *Moreover,* $R\phi \subset A$.

Proof. That $R\phi$ is a normal subgroup follows from the fact that $\phi^{\perp n}$ is an endomorphism for every n, and $G\phi^{\perp n} \subset G\phi^{\perp m}$ if $n \leq m$. We use a series of short steps for the remainder of the proof.

(a) $G\phi^{\perp m \perp n}$ is a group.

It is sufficient to show that $G\phi^n$ is a group for any crossed endomorphism ϕ since $\phi^{\perp m \perp}$ is a crossed endomorphism. If now $g, h \in G\phi^n$, then we have, with some inner automorphisms i, i', j, and j' whose nature we need not specify,

$$\phi^n(gh) = \phi^{n-1}(\phi(h)^{g^{-1}}\phi(g)) = \phi^{n-2}(\phi^2(g)^i \phi^2(h)^{i'}) = \cdots = \phi^n(g)^j \phi^n(h)^{j'} = 1.$$

That $G\phi^n$ is closed under forming inverses follows from 1.2.(iii).

(b) $G\phi^{\perp m \perp n} \subset G\phi^{\perp ms \perp nt}$.

If we set $\psi = \phi^{\perp m \perp}$, then ψ is a crossed endomorphism and $\psi^\perp = \phi^{\perp m}$, since $\perp$ is an involution. The assertion is now $G\psi^n \subset G\psi^{\perp s \perp nt}$. Let $g \in G\psi^n$. Now by 2.1.(iii) with some natural numbers m_i we have

$$\psi^{\perp s \perp nt}(g) = \prod_{k=1}^{s} \psi^{k+nt}(g)^{m_k} = 1.$$

(c) $C\phi$ is a group.

This follows now immediately from (b) since $G\phi^{\perp m \perp n} \cup G\phi^{\perp s \perp t} \subset G\phi^{\perp ms \perp nt}$.

(d) $C\phi \cap R\phi = \{1\}$.

We have to show that $G\phi^{\perp n \perp m} \cap G\phi^{\perp k} = \{1\}$ for all k, m, n. First we show that for a crossed endomorphism ϕ and all n we have $G\phi^n \cap R\phi = \{1\}$; because the sets $G\phi^n$ do not decrease, it is enough to show that $G\phi^n \cap G\phi^{\perp(n+m)} = \{1\}$ for $m = 1, \cdots$. Suppose that g is in that intersection, i.e.

$$\phi^n(g) = \phi^{\perp(n+m)}(g) = 1.$$

Then by 2.1.(iii) with $m' = n + m$

$$1 = \phi^{\perp m'}(g) = \prod_{k=1}^{n-1} \phi^k(g)^{(-1)^k (m';n)} g$$

since $\phi^{n+m}(g) = 1$, if $m \geq 1$. Therefore

$$g = \prod_{k=1}^{n-1} \phi^k(g)^{(-1)^{k-1}(m';n)}.$$

Now,

$$\phi^{n-1}(g) = \prod_{k=1}^{n-1} \phi^{n+k-1}(g)^{(-1)^{k-1}(m';n)} = 1.$$

By induction this implies $g = 1$.

We proceed now to the general case: $G\phi^{\perp n \perp m} \cap G\phi^{\perp k} = \{1\}$. If $k \leq n$, then we observe $G\phi^{\perp k} \subset G\phi^{\perp n}$; it is therefore sufficient to prove $G\phi^{\perp k \perp m} \cap G\phi^{\perp k} = \{1\}$

or $G\psi^m \cap G\psi^{\perp} = \{1\}$ for the crossed endomorphism $\psi = \phi^{\perp k \perp}$; but this is exactly what we proved above. If, however, $n \leq k$, then we remember that $G\phi^{\perp n \perp m} \subset G^{\perp nk \perp m}$ from section (b) of this proof; but this observation then leads us back to the case just discussed. With this remark we finish the proof of the proposition.

(e) $R\phi \subset A$.

If $g \in R\phi$, then $\phi^{\perp n}(g) = 1$ for some n. That is $\phi^{\perp}(\phi^{\perp n-1}(g)) = 1$. Hence, by induction on n, we need only show that $\phi^{\perp}(g) \in A$ implies $g \in A$. But actually these two relations are equivalent since $\phi(G) \subset A$ and $\phi^{\perp}(g) = \phi(g)^{-1}g$ for all $g \in G$.

Theorem II. *Let G be a group and ϕ a crossed endomorphism of G into an abelian subgroup A. Let $R\phi$ be the radical of the dual pair $(\phi, \phi^{\perp})$ and $C\phi$ its crossed radical. Then the following statements are true:*

(i) $R\phi$ is a normal subgroup of G and is contained in A,

(ii) $C\phi$ is a subgroup of G,

(iii) $R\phi \cap C\phi = \{1\}$,

(iv) $g \in R\phi \cdot C\phi$ if and only if there are integers $m, n,$ and s, such that
$$\phi^{\perp m \perp n}\phi^{\perp s}(g) = 1.$$

Proof. We have to prove (iv). Suppose first that $g = hk$ with $h \in R\phi$, $k \in C\phi$; then there are integers, $m, n,$ and s such that

$$1 = \phi^{\perp s}(h) = \phi^{\perp m \perp n}(k).$$

Thus $\phi^{\perp m \perp n}\phi^{\perp s}(g) = \phi^{\perp m \perp n}(\phi^{\perp s}(h)\phi^{\perp s}(k)) = \phi^{\perp m \perp n}(\phi^{\perp s}(k)) = \phi^{\perp s}(\phi^{\perp m \perp n}(k)) = 1$, because all powers of $\phi^{\perp}$ and ϕ commute since $\phi^{\perp}$ and ϕ commute (1.2.(v)).

To prove the converse we assume that $\phi^{\perp m \perp n}(\phi^{\perp s}(g)) = 1$ with some integers $m, n,$ and s. Since all powers of ϕ and $\phi^{\perp}$ commute, we have likewise $\phi^{\perp s}(\phi^{\perp m \perp n}(g)) = 1$.

We have to show that this assumption implies $g \in R\phi \cdot C\phi$. If $s < m$ then we apply $\phi^{\perp m-s}$ to the given identity and get $\phi^{\perp m \perp n} \cdot \phi^{\perp m}(g) = 1$. If $m < s$, we remember $G\phi^{\perp m \perp n} \subset G\phi^{\perp ms \perp n}$ (part d) of the proof to (2.3.) and derive $\phi^{\perp ms \perp n} \cdot \phi^{\perp s}(g) = 1$ from the given relation which in turn, according to the previous remark implies $\phi^{\perp ms \perp n} \cdot \phi^{\perp ms}(g) = 1$. After these observations it is sufficient to assume that we are given an identity of the form $\phi^{\perp m \perp n} \cdot \phi^{\perp m}(g) = 1$; we have to show that this implies $g \in R\phi \cdot C\phi$. We set $\psi = \phi^{\perp m \perp}$ so that $\psi^{\perp} = \phi^{\perp m}$. The given relation then takes the form $\psi^n \psi^{\perp}(g) = 1$. The radical of $(\psi, \psi^{\perp})$ is equal to $R\phi$ and the crossed radical of (ψ, ψ') is equal to $C\phi$ (as one can easily see from the definition and the fact that the groups $G\phi^{\perp n}$ respectively $G\phi^{\perp n \perp m}$

form increasing filter bases). Hence we have reduced this part of the proof to a proof of the following statement: If $\psi^n\psi^\perp(g)=1$, then $g\in R\psi\cdot C\psi$.

We shall prove this assertion by induction on n:

The statement is true for $n=1,\cdots,m$. By 1.2.(i), $\psi^{\perp m}(g)=g\psi^{\perp m\perp}(g)^{-1}$ or equivalently $g=\psi^{\perp m}(g)\cdot\psi^{\perp m\perp}(g)$ which by 2.1 implies

$$g=\prod_{n=1}^{m}\psi^n(g)^{(-1)^{n-1}(m;n)}\psi^{\perp m}(g).$$

Suppose now that $\psi^{m+1}\psi^\perp(g)=\psi^\perp\psi^{m+1}(g)=1$. Then

$$\psi^\perp\psi^{m+1-n}(\psi^n(g))=1 \text{ for } n=1,\cdots,m.$$

By induction hypothesis this implies $\psi^n(g)\in R\psi\cdot C\psi$ for $n=1,\cdots,m$. In order to show that g is actually in $R\psi\cdot C\psi$, it is now sufficient to establish that $\psi^{\perp m}(g)$ is in $C\psi$, because $R\psi$ is normal. But $\psi^{m+}(\psi^{\perp m}(g))=\psi^{\perp m-1}(\psi^\perp\psi^{m+1}(g))=1$, which shows indeed that $\psi^{\perp m}(g)$ is in the crossed radical. This observation finishes the proof.

III. Quotient Spaces with Invariant Means.

In this section again let G be a topological group (with the discrete topology admitted). We introduce a particular way of constructing crossed endomorphisms which split the group. Let N denote a closed subgroup and G/N the space of left cosets $x=gN$. If g is in G and $x=g'N\in G/N$, then gx will denote the left coset $gg'N$.

3.1. **Definition.** Let σ be a function from G/N into G satsifying the following conditions:

(i) σ is continuous,

(ii) $x=\sigma(x)N$.

Then σ is called a *cross section* for N in G. A *continuous cross section* is called *uniformly continuous* on a subspace S of G/N if besides (i) and (ii) the following condition is satisfied:

(iii) For each neighborhood U of 1 there is a neighborhood V of 1 in G such that

$$\sigma(Vx)\sigma(x)^{-1}\subset U \text{ for all } x\in S.$$

Note that, if G has the discrete topology, (i) and (iii) are automatically satisfied for every function σ whose range is a system of coset representatives.

3.2. Let σ be a continuous cross section for N in G. Then we define a function δ from $G\times G/N$ into G by

$$\delta(g,x) = g\sigma(x)\sigma(gx)^{-1}.$$

This function is called the *deviation*.

3.3 PROPOSITION.
(i) δ *is continuous*,
(ii) $\delta(gh,x) = \delta(h,x)^{g^{-1}}\delta(g,hx)$.

PROOF. (i) is trivial. (ii) We compute $gh\sigma(x)$. From the definition of the deviation we have $gh\sigma(x) = \delta(gh,x)\sigma(ghx)$ on one hand, and

$$gh\sigma(x) = g\delta(h,x)\sigma(hx) = \delta(h,x)^{g^{-1}}g\sigma(hx) = \delta(h,x)^{g^{-1}}\delta(g,hx)\sigma(ghx)$$

on the other. This proves (ii).

3.4. PROPOSITION. *If N is normal, then the range of the deviation δ is N, and $\delta(n,x) = n$ for all $n \in N$.*

PROOF. Since $\sigma(gx)$ and $g\sigma(x)$ are in the same coset gx, there is an $n' \in N$ such that $\sigma(gx)n' = g\sigma(x)$. Therefore $\delta(g,x) = \sigma(gx)n'\sigma(gx)^{-1} \in N$, because N is normal. Since N is normal, any coset $x = hN$ is equal to Nh. Thus for all $n \in N$, we have $nx = nhN = Nnh = Nh = x$. Hence, $\delta(n,x) = n\sigma(x)\sigma(nx)^{-1} = n$.

3.5. LEMMA. *Let $g \in G$. If σ is uniformly continuous on $gS \subset B/N$, then for each symmetric neighborhood U of 1 there is a neighborhood V of 1 such that*
(i) $\delta(Vg,x) \subset U\delta(g,x)$ *for all $x \in S$.*

PROOF. According to the definition of uniform continuity there is a neighborhood W of 1 such that $\delta(Wg,x) \subset Wg\sigma(x)\sigma(gx)^{-1}U = W\delta(g,x)U.$ We set $V = U \cap W$.

3.6. LEMMA. *If S is a compact subspace of G/N, then for each neighborhood U of 1 there is a neighborhood V of 1 such that $V\delta(g,x) \subset \delta(g,x)U$ for a fixed g and all $x \in S$.*

PROOF. Because of the continuity of the cross section and of the multiplication for any fixed $x \in G/N$ we find a neighborhood $V(x)$ of 1 and a neighborhood $W(x)$ of x in G/N such that $\delta(g,W(x))^{-1}V(x)\delta(g,W(x)) \subset U$. Since S is compact, there is a finite number of neighborhoods $W(x_i)$, $i = 1,$

$\cdots, n,$ whose union covers S. We let V be the intersection of $V(x_i)$, $i = 1,$ $\cdots, n$. Then clearly for all $x \in S$ we have $\delta(g,x)^{-1} V \delta(g,x) \subset U$ which proves the assertion.

3.7. PROPOSITION. (i) *If S is a compact set in G/N, then the collection of functions $g \to \delta(g,x)$ of G into G with $x \in S$ is equicontinuous at 1.*

(ii) *If the collection Δ of mappings $\chi_g : G/N \to G$ defined by $\chi_g(x) = \delta(g,x)$ is given the compact open topology (i.e. the uniform convergence on compact subsets), then the mapping $g \to \chi_g$ of G into Δ is continuous.*

PROOF. Let S be a compact subset of G/N. We shall show that for each $g \in G$ and each neighborhood U of 1 there is a neighborhood V of 1 in G such that $\delta(Vg,x) \subset \delta(g,x) U$ uniformly. This will prove both (i) and (ii). Let U' be a neighborhood of 1 such that $U'^2 \subset U$. By 3.6. for each $g \in G$ there is a $W \subset U'$ such that $W\delta(g,x) \subset \delta(g,x) U'$ for all $x \in S$. Since S is compact, σ is uniformly continuous on S. Therefore, according to 3.5., there is a V such that

$$\delta(Vg,x) \subset W\delta(x,g)\, W, \text{ from which we now get}$$

$$\delta(Vg,x) \subset \delta(g,x)\, U'\, W \subset \delta(g,x)\, U'^2 \subset \delta(g,x)\, U.$$

3.8. DEFINITION. Let G be a topological group and N a closed normal subgroup, Δ a family of mappings from G/N into N with the topology of uniform convergence on compact sets. Then the pair (G,N) is said to admit an invariant mean M on Δ if there is a mapping M of Δ into B, whose operation will be described as $M(\chi)$, or equivalently as $M_x(\chi(x))$, which satisfies the following conditions:

(i) M is continuous,

(ii) $M_x(\chi(x)\chi'(x)) = M_x(\chi(x))M_x(\chi'(x))$,

(iii) M commutes with all automorphisms of N, i.e. $M_x(\chi(x)^\alpha) = M_x(\chi(x))$ for any automorphism α of N,

(iv) $M_x(\chi(gx)) = M_x(\chi(x))$ (invariance),

(v) $M_x(n) = n$ for the constant functions $n(x) = n$ with $n \in N$.

If a functional M from Δ to N satisfies (i)-(iv), then M will be called an *invariant pro-mean* on Δ.

3.9. PROPOSITION. *Let G be a topological group and N a closed normal subgroup admitting a cross section. If (G,N) admits an invariant pro-mean M on a space of mappings Δ containing the functions $x \to \delta(g,x)$ for all $g \in G$,* *then the mapping defined by*

$$\phi(g) = M_x(\delta(g,x))$$

is a crossed endomorphism of G into N. If, moreover, M is a mean, then $N = G\phi^\perp = \phi(G)$.

PROOF. (a) ϕ is continuous because of 3.7. and the continuity of M.
(b) By 3.3.(ii) we have

$$\phi(gh) = M_x(\delta(h,x)^{g^{-1}}\delta(g,x)) = M_x(\delta(h,x)^{g^{-1}})M_x(\delta(g,x)) =$$
$$= M_x(\delta(h,x))^{g^{-1}}M_x(\delta(g,x)) = \phi(h)^{g^{-1}}\phi(g)$$

using the definition of a pro-mean. This proves that ϕ is a crossed endomorphism of G into N.

(c) Let now M be a mean. Then for $n \in N$ we have $\phi(n) = M_x(n) = n$ which shows that N is contained in the fixed point set ϕG of ϕ. Hence $\phi(G) \subset N \subset \phi G = G\phi^\perp$. But $G\phi^\perp \subset \phi(G)$ by 1.5.(ii), which finishes the proof.

3.10. PROPOSITION. *Let G be a topological group which can be represented in two (possibly different) ways as a semidirect product $G = NC = ND$ of a normal closed subgroup N and closed subgroups C and D. Let $\pi:G/N \to C$ and $\rho:G/N \to D$ be the obvious isomorphisms from G/N to C and D respectively. Then the mapping $x \to \pi(x)\rho(x)^{-1}$ is a continuous mapping from G/N to N; and if (G,N) admits an invariant mean for a function space containing this mapping, then C and D are conjugate.*

PROOF. Since $\pi(x)$ and $\rho(x)$ are in the same coset modulo N, the mapping $\chi:x \to \pi(x)\rho(x)^{-1}$ maps G/N not only into G but in fact into N. The functional equation

$$\chi(xy) = \chi(y)^{\pi(x)^{-1}}\chi(x)$$

is easily computed:

$$\chi(xy) = \pi(x)\pi(y)\rho(y)^{-1}\rho(x)^{-1} = \pi(x)\pi(y)\rho(y)^{-1}\pi(x)^{-1}\pi(x)\pi(x)^{-1} = \chi(y)^{\pi(x)^{-1}}\chi(x).$$

The similarity to the defining identity for a crossed endomorphism is obvious. Now we let $M_y(\chi(xy)) = M_y(\chi(y)) = n$ and apply the invariant mean to both sides of the functional equation. Then we get

$$n = M_y(\chi(y)^{\pi(x)^{-1}}) M_y(\chi(x)) = n^{\pi(x)^{-1}} \chi(x) = \pi(x) n \pi(x)^{-1} \pi(x) \rho(x)^{-1}.$$

This implies $\rho(x) = n^{-1}\pi(x)n = \pi(x)^n$, which finishes the proposition.

THEOREM IV. *Let G be a topological group and N a closed normal subgroup admitting a cross section. If (G,N) admits an invariant mean for a function space containing all mappings $x \to \delta(g,x)$, $g \in G$, then N splits in G, i.e. there is a closed subgroup C of G such that*

(i) $G = NC$,

(ii) $N \cap C = 1$,

(iii) *C is isomorphic to G/N and $(n,c) \to nc$ is a homeomorphism of $N \times C$ onto G.*

If $G = NC = ND$, and these products are semidirect, if, furthermore, (G,N) admits an invariant mean for $x \to \pi(x)\rho(x)^{-1}$ (π and ρ as in 3.10), then C and D are conjugate.

PROOF. The crossed endomorphism ϕ defined in the previous proposition satisfies condition (iv) of Theorem I and $N = \phi(G)$. Hence Theorem I applies and proves the first part of the assertion. The second is a consequence of 3.10.

REMARK. Throughout this work, the normality of N could be replaced by the hypothesis that the range of δ is N, which then, by the Theorem, implies that N is normal.

THEOREM IV LOC. *Let G be a topological group and N a closed normal subgroup admitting a cross section. If (G,N) admits an invariant mean for a space Δ of mappings containing the functions $x \to \delta(g,x)$ for all g in a neighborhood U of the unit in G, then N splits locally, i.e. there exists a closed local subgroup C_l such that*

(i) *NC_l is a neighborhood of 1,*

(ii) $N \cap C_l = 1$,

(iii) *$(n,c) \to nc$ is a homeomorphism of $N \times C_l$ onto NC_l.*

PROOF. We set $\phi(g) = M_x(\delta(g,x))$ for all $g \in U$. Similarly to our procedure in 3.8., we see that ϕ is a local crossed endomorphism. Furthermore $\phi(n) = n$ for all $n \in N \cap U$, which implies $\phi^2(g) = \phi(g)$ for sufficiently small g so that condition (i) of Theorem I loc is satisfied. Let $C_l = U\phi$. Then (i) and (ii) are proved by Theorem I loc. The mapping $(n,c) \to nc$ is continuous and one-one. By Theorem I loc., for sufficiently small neighborhoods V of 1, the mapping $(n,c) \to nc$ of $V\phi^{\perp} \times V\phi \to V\phi^{\perp} \cdot V\phi$ is a homeomorphism. Then the sets $nV\phi^{\perp}$

and$/gV_\phi, n\in N$ and $g\in C_l$ form bases for the topologies on N and C_l respectively. Hence $N\times C_l\mapsto NU$ is indeed a homeomorphism.

IV. First Application.

As in section II, no topology enters the considerations of this section. We shall be concerned with an abstract group G and an abelian normal subgroup A with finite index $(G:A)=n$. The results of this section are more or less well-known.

4.1. Proposition. *The pair (G,A) admits an invariant pro-mean on the set of all functions from G/A to A.*

Proof. For each function $\chi:G/A\to A$ we set

$$M_x(\chi(x))=\prod\{\chi(y):y\in G/A\}.$$

Conditions (i)–(iv) of Definition 3.8. are easily checked.

4.2. Proposition. *There exists a crossed endomorphism ϕ of G into A such that $\phi(a)=a^n$ and $\phi^\perp(a)=a^{1-n}$, for $a\in A$. Moreover, $G=G\phi^\perp$ is the set of all points in A whose order divides $n-1$.*

Proof. Let $\sigma:G/A\to A$ be a cross section which trivially exists because there are only finitely many cosets modulo A. If we define ϕ as in Proposition 3.9., then it satisfies the requirements: By Proposition 3.9. it is a crossed endomorphism. If $a\in A$, then, by 3.4., $\delta(a,x)=a$ independently of x. Therefore $\phi(a)=a^n$, which, by 1.2. (ii) implies $\phi^\perp(a)=a^{1-n}$. A point g of G is a fixed point of ϕ if, as an image under ϕ, it is in A and satisfies the identity

$$\phi(a^n)=a.$$

Theorem V. *Let G be a group with an abelian normal subgroup A of finite index which is uniquely divisible by its index n (i.e. the equation $x^n=a$ is uniquely solvable in A). Then G is a semidirect product of A and a finite group C of order n. If G is also a semidirect product of A and D, then C and D are conjugate.*

Proof. The restriction of the crossed endomorphism of Proposition 4.2. to A is an isomorphism into A, and $\phi(a)=a^n$ clearly is permutable with all inner automorphisms of G. Then the first part of the Theorem follows by the Corollary to Theorem I. The conjugacy of C and D follows from Theorem IV, since (G,N) admits the invariant mean M defined by

$$M_x(f(x)) = n^{-1}(f(x_1) + \cdots + f(x_n)), \quad \{x_1, \cdots, x_n\} = G/A.$$

COROLLARY V.1. (SCHUR). *An abelian normal torsion subgroup with finite index splits if its index is relatively prime to all the orders of its elements. (See, for instance, [24, p. 132].)*

PROOF. Under these hypotheses A is uniquely divisible by its index.

4.3. **DEFINITION.** The set of all primes will be denoted by Γ. If n is a natural number, then $\Gamma(n)$ is the set of all primes which divide n. If A is an abelian group, Σ a set of primes, then A_Σ will denote the subgroup of all elements whose order is a product of powers of primes in Σ. Remember that for any abelian torsion group one has a direct decomposition of A into A_Σ and $A_{\Gamma \setminus \Sigma}$.

THEOREM VI. *Let G be a group with an abelian normal torsion subgroup A of finite index n. Let m be a number relatively prime to n. Then G is the semidirect product of a normal subgroup N and a subgroup C such that $N = A_{\Gamma(m)}$, $C \cap A = A_{\Gamma \setminus \Gamma(m)}$, and $C \cap A$ is normal in G.*

REMARK. This decomposition is in a sense an extension of the direct decomposition of A into $A_{\Gamma(m)} \cdot A_{\Gamma \setminus \Gamma(m)}$.

PROOF. Let $N = A_{\Gamma(m)}$ and $C' = A_{\Gamma / \Gamma(m)}$; since C' is characteristic in A it is normal in G. The group $\overline{G} = G/C'$ satisfies the hypotheses of Corollary V.1. Therefore $\overline{G}$ is a semidirect product of A/C' and some subgroup C/C'. Then $G = AC$ and $A \cap C = C'$; since C contains C' and $A = NC'$, the identity $G = AC$ implies $G = NC$; from $A \cap C = C'$ we get $N \cap C = \{1\}$. This finishes the proof.

REMARK. This theorem can also be proved using Proposition 4.2. and Theorem II.

<h3 align="center">V. SPLITTING OF VECTOR SUBGROUPS WITH COMPACT QUOTIENT.</h3>

5.1. **DEFINITION.** Let G be a topological group and N a closed subgroup such that the quotient space G/N of left cosets is a compact space. Let B be a reflexive Banach space and $C(G/N,B)$ the vector space of all continuous functions of G/N into B with the topology of uniform convergence. A continuous linear functional I on $C(G/N,B)$ is called an *invariant integral* and is denoted by $I(\phi) = I_x(\phi(x)) = \int \phi(x)dx$ if it satisfies

$$I_x(\phi(gx)) = I_x(\phi(x)) \text{ for all } g \in G.$$

An integral is called *normed* if it assigns to constant functions their constant value.

5.2. PROPOSITION. *If there is an invariant normed integral for $C(G/N,B)$ when B is the reals, then there is an invariant normed integral for $C(G/N,B)$ when B is an arbitrary reflexive Banach space.*

PROOF. Let $(\phi,\phi') \to \langle \phi,\phi' \rangle$ be the bilinear form on $B \times B'$ defining the duality between B and B', the dual of B. Then, for all $\phi' \in B'$ and a fixed function ϕ from $C(G/N,B)$, the mapping $x \to \langle \phi(x), \phi' \rangle$ is a continuous real-valued function. Using the invariant integral $f \to \int f(x)d_hx$ for real-valued functions defined on the compact group G/N having norm $\int 1d_hx = 1$, we define a linear functional on B' by

$$\phi' \to \int \langle \phi(x), \phi' \rangle d_hx.$$

This functional is bounded, because

$$\left| \int \langle \phi(x), \phi' \rangle d_hx \right| \leq \int |\langle \phi(x), \phi' \rangle| d_hx \leq \|\phi'\| \int \|\phi(x)\| d_hx.$$

Since B' is reflexive, there is an element $I(\phi)$ in B representing this functional. In other words, $I(\phi)$ is defined by

$$\langle I(\phi), \phi' \rangle = \int \langle \phi(x), \phi' \rangle d_hx.$$

It is easy to see that $\phi \to I(\phi)$ is a linear form. It is continuous; for

$$|\langle I(\phi), \phi' \rangle| = \left| \int \langle \phi(x), \phi' \rangle d_hx \right| \leq \int |\langle \phi(x), \phi' \rangle| d_hx \leq \int \|\phi(x)\| \|\phi'\| d_hx$$

$$\leq \|\phi'\| \max \{\|\phi(x)\| : x \in G/N\};$$

by noting that the norm $\|\phi\|^*$ of ϕ in $C(G/N,B)$ is $\max\{\|\phi(x)\| : x \in G/N\}$ we get $|\langle I(\phi), \phi' \rangle| \leq \|\phi\|^* \cdot \|\phi'\|$; therefore $\|I(\phi)\| \leq \|\phi\|^*$ from which we get $\|I\| \leq 1$. Thus I is an integral. Because of the invariance of the Haar integral it is obviously invariant. If $\phi(x) = b$ is a constant function, then $\langle I(\phi), \phi' \rangle =$

$\int \langle b, \ \phi' \rangle d_h x = \langle b, \phi' \rangle$, which shows $I(\phi) = b$. It is clear that the functional I can be extended to the space of bounded measurable functions.

5.3. PROPOSITION. *If, in addition to the assumptions of 5.1., N is normal, then there exists an invariant integral.*

PROOF. G/N is a compact group and admits an invariant integral for real-valued functions. Then the proposition follows from 5.2.

5.4. PROPOSITION. *If, in addition to the assumptions of 5.1., G is locally compact and the unimodularity factor $\Delta(g)$ of G coincides on N, with the unimodularity factor $\delta(g)$ of N then there exists an invariant integral. (The unimodularity factor $\Delta(g)$ on G is defined by $\int f(gh^{-1})dg = \Delta(g) \int f(g)dg$ for a left invariant Haar integral* [23, p.39].)

PROOF. It is proved in [23, p. 45] that under these hypotheses there exists an invariant integral for $C(G/N,B)$ if B is one dimensional.

THEOREM VII. *Let G be a topological group and N a closed normal subgroup satisfying the following conditions:*
   (i) *N is isomorphic to the additive group of a reflexive Banach space,*
   (ii) *G/N is compact,*
   (iii) *N admits a cross section in G.*
Then N splits in G, i.e. there is a compact subgroup C in G isomorphic algebraically and topologically to G/N such that G is the semidirect product of N and C. If $G = NC = ND$ and these products are semidirect, then C and D are conjugate.

REMARK. It is sufficient to require the existence of a local cross section instead of (iii) [20, p.55].

PROOF. We prove that (G,N) admits an invariant mean for all continuous functions $x \rightarrow \phi(x)$ of G/N into N. Then Theorem IV will finish the proof. Let χ be a continuous and continuously invertible isomorphism of N onto the reflexive Banach space B. Let I be the invariant integral for $C(G/N,B)$, which exists from 5.3. Then we define

$$M_x(\phi(x)) = \chi^{-1}I_x(\chi(\phi(x))).$$

It is clear that the conditions (i), (ii), (iv), and (v) of 3.8. are satisfied. Condition (iii) is equivalent to $I_x(\psi(x)^a) = I(\psi)^a$ for all continuous automorphisms of B; but this condition is satisfied because

$$\langle I_x(\psi(x)^\alpha),\ \psi'\rangle = \int\langle\psi(x)^\alpha,\ \psi'\rangle d_h x = \int\langle\psi(x),\ \psi'^{\alpha'}\rangle d_h x$$
$$\langle I_x(\psi(x)),\ \psi'^\alpha\rangle = \langle I_x(\psi)^\alpha,\ \psi'\rangle$$

with the adjoint α' of α. We should like to remind the reader in this context of Example 1.7.

THEOREM VII LOC. *Let G be a topological group and N a closed normal subgroup satisfying the following conditions:*

(i) *N is locally isomorphic to a reflexive Banach space B (i.e. the connected component of N has B as its simply connected covering group),*

(ii) *G/N is compact,*

(iii) *N admits a continuous cross section in G. Then N splits locally, i.e. there exists a compact local group C' such that*

(i') *NC' is a neighborhood of 1,*

(ii') *$N\cap C' = \{1\}$,*

(iii') *$(n,c)\to nc$ is a homeomorphism of $N\times C'$ onto NC'.*

PROOF. We prove that (G,N) admits an invariant mean for a set of functions containing all functions $x\to\delta(g,x)$ for sufficiently small g. Then Theorem IV loc. will finish the proof. By 3.7.(i) the mappings $g\to\delta(g,x)$ are equicontinuous at 1. Therefore there is a neighborhood V of 1 in G such that $\delta(g,x)$ is in a preassigned neighborhood for all $g\in V$, $x\in G/N$. Let now χ be a local isomorphism mapping a neighborhood W of 1 in N onto a neighborhood of 0 in a reflexive Banach space B. Let I be the invariant integral on $C(G/N,B)$. We define M by setting

$$M_x(\phi(x)) = \chi^{-1} I_x(\chi\phi(x)).$$

Then M is an invariant mean for all continuous functions $x\to\phi(x)$ with $\phi(G/N)\subset W$. This collection contains in particular all functions $x\to\delta(g,x)$ for $g\in V$.

VI. SECOND APPLICATION.

6.1. PROPOSITION. *If G is a metric topological group and N a closed subgroup isomorphic to a Banach space, then N admits a continuous cross section. If N is locally isomorphic to a Banach space, then N admits a local cross section.*

PROOF. This is a consequence of a theorem by Michael [14].

COROLLARY VII. 1. *Let G be a topological group satisfying the first axiom of countability, and let N be a closed normal subgroup. Let the following conditions be satisfied:*
  (i) *N is isomorphic to the additive group of a reflexive Banach space,*
  (ii) *G/N is compact.*
Then there is a compact subgroup C such that G is the semidirect product of N and C.
  If $G = NC = ND$ and these products are semidirect then C and D are conjugate.

PROOF. This follows from Theorem VII and 6.1.

VII. NORMAL HILBERT SUBGROUPS IN TOPOLOGICAL GROUPS.

7.1. LEMMA. *Let G be a topological group, F a closed normal subgroup and N a closed normal subgroup whose centralizer contains F. Let $p: G \to G/F$ be the coset homomorphism, $i(g)$ be the automorphism $n \to g^{-1}ng$ of N and let j be the identity of N into G. Then the set $i(p^{-1}(x)) \circ j$ contains one and only one automorphism $I(x)$ of N. If $\mathfrak{A}$ is the group of automorphisms of N in the compact open topology, then $x \to I(x)$ is a continuous mapping of G/F into $\mathfrak{A}$. If the centralizer of N is equal to F, then G/F acts effectively on N.*

PROOF. $g \to i(g) \circ j$ is an algebraic homomorphism of G into $\mathfrak{A}$. Let K be a compact subspace of G and U an open neighborhood of 1 in G. We prove that there is a neighborhood V of 1 in G such that $g \in V$ implies $k^{i(g)} \in kU$ for $k \in K$, which will prove that $g \to i(g) \circ j$ is a representation of G into $\mathfrak{A}$ if one considers on $\mathfrak{A}$ the topology of uniform convergence on compact sets. So we have to show that $g^{-1}kg \in kU$ or $k^{-1}g^{-1}kg \in U$ for $g \in V$, $k \in K$. Because of the continuity of multiplication, for each $k \in K$ there exists a neighborhood W_k of k in G and a neighborhood V_k of 1 in G so that $W_k^{-1}V_k^{-1}W_kV_k \subset U$. Since K is compact, a finite number $W_{k_1}, \cdots, W_{k_n}$ of neighborhoods cover K. If we put $V = V_{k_1} \cap \cdots \cap V_{k_n}$, then the requirement for V is fulfilled.

Since F is in the centralizer of N, the kernel of the homomorphism $g \to i(g) \circ j$ contains F; therefore $g'g^{-1} \in F$ implies $i(g') \circ j = i(g) \circ j$. Thus, for a coset $x \in G/F$, the set $i(p^{-1}(x)) \circ j$ contains precisely one automorphism $I(x)$ of $\mathfrak{A}$ and we have $I \circ p = i$. Since i is continuous, the following diagram shows that I is continuous:

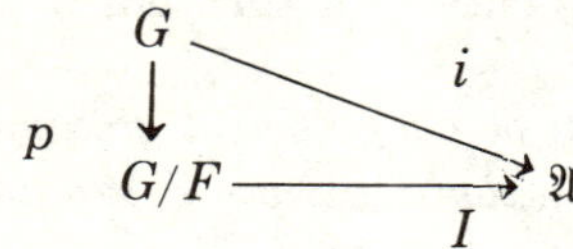

Now clearly the elements of G leaving N pointwise fixed are precisely those elements of the centralizer of N.

7.2. LEMMA. *If H is a Hilbert space and $\mathfrak{A}$ a group of invertible operators which is compact in the strong operator topology, then there exists a new inner product on H defining the same topology, such that $\mathfrak{A}$ is a unitary group relatively to this inner product.*

PROOF. This is a well-known fact [18, p.431].

7.3. LEMMA. *If H is a Hilbert space and $x \rightarrow I(x)$ is a representation of a compact group C into the group of unitary operators, then H splits into a direct orthogonal sum of finite dimensional subspaces invariant under $I(C)$.*

PROOF. See e.g. [13, p. 434].

7.4. PROPOSITION. *Let G be a topological group and N a closed normal subgroup isomorphic (algebraically and topologically) to the additive group of a Hilbert space, such that G/N is compact. Then there is a collection $\{N_j : j \in J\}$ of closed normal subgroups of G such that*

(i) N_j *is a finite dimensional vector group for all $j \in J$,*

(ii) $N_j \cap N_{j'} = \{1\}$ *for $j \neq j'$,*

(iii) N *is the product of the N_j in the following sense:*

If χ is an isomorphism of N onto a Hilbert space H, then H is the Hilbert space sum of the subspaces $\chi(N_j)$.

PROOF. After 7.1. G/N acts as a group of automorphisms on N which is compact in the compact open topology. Therefore G/N is represented continuously into the group of invertible operators of a Hilbert space H being an isomorphic image of N under an isomorphism χ. The compact open topology is finer than the strong operator topology which is nothing else than the point open topology. Hence, by Lemmas 7.2. and 7.3., N has the structure described in the theorem.

In the following, G shall always denote a topological group with a closed normal subgroup N such that N is isomorphic to the additive group of a

Hilbert space and G/N is compact. Without assuming the existence of a cross section, we are going to construct a closed subgroup C in G such that $G=NC$, $N\cap C=\{1\}$. More specifically, we shall show that there is a one to one representation $f:G\to G'$ of G onto a topological group G' such that G' is the semidirect product of $N'=f(N)$ and $C'=f(C)$. An appropriate example will show that the proposed procedure, in general, does not lead to a semidirect decomposition of G. We are unable to decide whether or not it does in the case we consider.

Let $\mathfrak{M}$ denote the set of all normal subgroups H of G contained in N such that the quotient N/H is a finite dimensional vector space; G' shall denote the projective limit of the groups G/H, $H\in\mathfrak{M}$ and $f:G\to G'$ the obvious representation. Let, furthermore, $f':G/N\to G'/N'$ be the mapping which assigns to a coset $gN\in G/N$ the coset $f(g)N'\in G'/N'$.

7.5. PROPOSITION.

(i) *f is one to one,*

(ii) *f' is an isomorphism,*

(iii) *N' is isomorphic to the additive group of a Hilbert space endowed with the coarsest topology making all linear functionals of a certain separating set continuous.*

PROOF.

(i) By 7.4., the intersection of $\mathfrak{M}$ is the identity; this implies (i).

(ii) Trivially, f' is an algebraic isomorphism. We shall show that it is a representation, i.e. that it is continuous; then it will turn out to be an isomorphism because G/N is compact.

Let W be a neighborhood of the identity in G'/N'. By the definition of the quotient topology, W contains a neighborhood of the form $V'N'/N'$ with some neighborhood V' of the identity in G'. According to the definition of the projective limit, (see [23, p. 23–29]) V' contains a neighborhood $f(UH)$ where U is some neighborhood of 1 in G and H some subgroup of N. Now W contains $f(UH)N'/N'=f(U)N'/N'$. Hence $f'(UN/N)=f(U)N'/N'\subset W$; but UN/N is a neighborhood of the identity in G/N. This proves that f' is indeed continuous.

(iii) The neighborhood filter of the identity in N' is generated by the sets $f(UH)$, where U is a neighborhood of 1 in G and $H\in\mathfrak{M}$. For the moment, we may identify N' with the additive group of a Hilbert space; $f(H)$ is then a linear subspace of finite codimension and is, therefore the intersection of a finite number of hyperplanes which, in turn, are determined by a set of linear functionals. The sets $f(U)$ are neighborhoods of zero in the strong topology; the topology generated by the sets $f(U)f(H)$ is just fine enough to render the collection of all the linear functionals continuous which belong, in the above sense, to the co-finite dimensional subspaces $f(H)$, $H \in \mathfrak{M}$, and no coarser topology will do the same service. Finally we observe that this set of linear functionals separates the points of N'; but this again is a consequence of the fact that the intersection of all $f(H)$, $H \in \mathfrak{M}$ is the zero of N'.

7.6. PROPOSITION. (*Iwasawa*). *If G is locally compact then G is the semi-direct product of N and some compact subgroup C.*

PROOF. In this case, N has a cross section in G ([16] and [20, p. 55]); the assertion then follows from Theorem VII.

7.7. PROPOSITION. *$f:G \to G'$ is an isomorphism if and only if G is locally compact.*

PROOF. If G is locally compact then N is locally compact and, therefore, a finite dimensional vector group. The projective limit of the groups G/H, $H \in \mathfrak{M}$ is isomorphic to the projective limit of the groups G/H, $H \in \mathfrak{N}$ where $\mathfrak{N}$ is any cofinal subset of $\mathfrak{M}$ [23, p. 24]. If N is finite dimensional the set $\mathfrak{N} = \{\{1\}\}$ is such a cofinal subset of $\mathfrak{M}$. Hence G' is isomorphic to G.

Conversely, if G is not locally compact then N is not locally compact and hence (up to isomorphy) an infinite dimensional Hilbert space. In this case, by 7.5.(iii), the restriction of f to N is not an isomorphism; therefore f can not be an isomorphism.

7.8. PROPOSITION. *G contains a closed subgroup C which has the following properties:*
(i) $G = NC$,
(ii) $N \cap C = \{1\}$,
(iii) *G' is the semidirect product of N' and $C' = f(C)$.*

PROOF. Let $H \to H_c$ be a mapping from a directed subset $\mathfrak{N}$ of $\mathfrak{M}$ into the set of all closed subgroups of G such that
(a) $G = NH_c$,
(b) $N \cap H_c = H$,
(c) H_c/H is compact,
(d) $H < K$, H, $K \in \mathfrak{N}$ implies $H_c < K_c$.
The set of all these mappings is partially ordered by inclusion (we note

that mappings are sets). It is clearly inductive. Let now $H\to H_c$ be a maximal mapping of this type and $\mathfrak{N}$ its domain. We shall prove that $\mathfrak{N}$ is cofinal in $\mathfrak{M}$. Suppose not. Then there is an H such that no element of $\mathfrak{M}$ smaller than H is in $\mathfrak{N}$. We pick a $K<H$, $K\in\mathfrak{M}$. Let, for the moment, $\overline{G}=G/K$, $\overline{N}=N/K$ and $\overline{D}=H_c/K$. Then we have

(a') $\overline{G}=\overline{N}\overline{D}$,

(b') $\overline{N}\cap\overline{D}=\overline{H}=H/K$.

By 7.6, $\overline{D}$ is the semidirect product of $\overline{H}$ and some compact subgroup $\overline{C}_o$ since $\overline{D}$ is a locally compact group with a closed normal vector subgroup $\overline{H}$ such that $\overline{D}/\overline{H}$ is compact. Let K_c be the full counterimage of $\overline{C}_o$ in G. The mapping $H\to H_c$ is thus extended to $\mathfrak{N}\cup\{K\}$; the conditions (a), (b), (c) above are automatically satisfied. In order that (d) is satisfied, too, it is necessary and sufficient that for all $L\in\mathfrak{N}$, $K<L$ the group $\overline{C}_o$ is contained in all L_c/K; by the induction hypothesis, these form a set which is directed by inclusion; since G/K is locally compact, N/K is a finite dimensional vector space and all chains in this directed set are finite; therefore there are minimal elements, but all these have to coincide because the set is directed. The unique minimal element is necessarily $\overline{D}=H_c/K$ which contains $\overline{C}_o$ by the definition of $\overline{C}_o$. Hence condition (d) is satisfied, too. From now on we may replace the set $\mathfrak{M}$ by the cofinal set $\mathfrak{N}$.

Let now C' be the subgroup of all $(gH)_{H\in\mathfrak{N}}\in G'$ such that $gH\in H_c/H$ for all $H\in\mathfrak{N}$. Clearly C' is closed in G'. Let $C=f^{-1}(C')$. For all $K\in\mathfrak{N}$ we have

$$CK/K=\bigcap\{H_cK/K: H\in\mathfrak{N}, H<K\}.$$

Condition (a) above implies $G/K=N/K\cdot H_cK/K$ for all $H\in\mathfrak{N}$; by (c), H_c/H is compact, therefore $(H_c/H)(K/H)/(K/H)$ is isomorphic to H_c/H for all $H\in\mathfrak{N}$, $H<K$; by the isomorphy theorem, $(H_cK/H)/(K/H)$ and H_cK/K are isomorphic; this shows that H_cK/K is compact. The vector group N/K, however, does not contain any nontrivial compact subgroup; hence $N/K\cap H_cK/K=\{K/K\}$, and since H_cK/K is compact, the product $N/K\cdot H_cK/K$ is semidirect. Let now $H,J\in\mathfrak{N}$, $H,J<K$. Then there is an $L\in\mathfrak{N}$, $L\subset H\cap J$. The groups H_cK/K, J_cK/K, and L_cK/K are isomorphic to $(G/K)/(N/K)$ under the obvious mappings. But L_cK/K is contained in the other two; therefore all three must coincide. This shows that actually

$$CK/K=H_cK/K \quad \text{for all} \quad H\in\mathfrak{N}, H<K.$$

Thus, for all $K\in\mathfrak{N}$, the group G/K is the semidirect product of N/K and

CK/K. Therefore $G=NC$ and $N\cap C=\{1\}$. Furthermore, $C'=f(C)$ is the projective limit of the compact groups CK/K, $K\in\mathfrak{N}$ and is, consequently, compact. The representation of C' onto G'/N' is now an isomorphism, and the product $N'C'$ is semidirect.

There is some evidence that the restriction of f to C is, under certain circumstances, actually an isomorphism. The fact that N is a vector space would have to be used strongly; under slightly more general conditions, this is certainly no longer true: Let G be the group of our Example 1.7. Let $\mathfrak{M}$ be the collection of all subgroups H of N consisting of elements $((x_i)_{i\in I}, 1)$ such that $x_i=1$ for a fixed finite set of indices. Let G' be the projective limit of the groups G/H, $H\in\mathfrak{M}$. Now G/H is the direct product of N/H which is a finite dimensional torus group and of CH/H which is an infinite dimensional compact torus group; therefore both G' and $C'=f(C)$ are compact where f again denotes the one to one representation of G onto G'. Since C is not compact the restriction of f to C is not an isomorphism. We are now ready to formulate

THEOREM VIII. *Let G be a topological group and N a closed normal subgroup. Let the following two conditions be satisfied:*

(i) *G/N is compact,*

(ii) *N is isomorphic to the additive group of a Hilbert space.*

Then the following statements are valid:

(i') *There is a closed subgroup C such that $G=NC$, $N\cap C=\{1\}$.*

(ii') *There is a collection $\mathfrak{M}$ of subgroups of N directed by inclusion which are normal in G such that the obvious representation f of G onto the projective limit G' of the groups G/H, $H\in\mathfrak{M}$ is one to one; moreover, all groups G/H, $H\in\mathfrak{M}$ are locally compact and G' is the semidirect product of $f(N)$ and $f(C)$. The representation f is an isomorphism if and only if G is locally compact. G/N, $G'/f(N)$ and $f(C)$ are isomorphic.*

(iii') *If N admits a cross section in G (which is true e.g. in the case that G is further locally compact or satisfies the first axiom of countability) then G is the semidirect product of N and some compact subgroup of G.*

REMARK. We do not say that f/C is an isomorphism if and only if G is locally compact. It is possible that f/C is always an isomorphism.

VIII. SPLITTING OF VECTOR SOLVABLE SUBGROUPS.

8.1. DEFINITION. Let G be a topological group. Then the commutator group G' of G is the closure of the subgroup generated by all commutators of G. The group G is called *solvable* if the series

$$G \supset G' \supset G^{(2)} \supset G^{(n-1)} \supset G^{(n)} = \{1\}$$

terminates after a finite number of steps, where obviously $G^{(m+1)} = (G^{(m)})'$ for $m = 1, \cdots$. We call a topological group *vector solvable* if it is solvable and the factor groups $G^{(i)}/G^{(i+1)}$, $G^{(0)} = G$, $i = 0, \cdots, n-1$ are isomorphic to the additive group of reflexive Banach spaces H_i. The natural number n is called the length of G.

THEOREM IX. *Let G be a topological group with a closed normal subgroup N which is vector solvable. Suppose that G/N is compact and that G is locally compact or first countable. Then G is a semidirect product of N and a compact subgroup C. If $G = NC = ND$ and these products are semidirect, then C and D are conjugate.*

PROOF. We prove the theorem by induction on the length of N. If the length of N is one, the theorem is true by Theorem VIII. Suppose that the theorem is true for solvable subgroups of length $n-1$ and suppose that N has length n, i.e. there is a decreasing sequence of commutator groups $N = N^{(0)} \supset \cdots \supset N^{(n)} = \{1\}$. The closed subgroup $M = N^{(n-1)}$ is isomorphic to the additive group of a Hilbert space; it is characteristic in N, hence normal in G. The group G/M contains a normal subgroup N/M which satisfies the hypotheses made about N in the theorem; it is, however, of length $n-1$ so that the induction hypothesis applies and yields a compact subgroup C_o/M. But C_o satisfies the hypotheses of Theorem VIII; therefore it contains a compact subgroup C such that it is the semidirect product of M and C. Hence, $G = NMC = NC$ since M is contained in N. Let now $G = NC = ND$; then, by induction, CM/M is a conjugate of DM/M in G/M, i.e. there is an $n \in N$ such that $C^n \subset DM$. Now we have $MC^n = MD$, and each of these products is semidirect, since C^n and D are compact, and $C^n \cap M = D \cap M = \{1\}$. Then C^n and D are conjugate in $C^n M$ after Theorem IV, hence, a fortiori, in G.

REMARK. If G is locally compact, then G is in particular a C-group in the sense of Iwasawa [9], (see also 12.11). The hypothesis implies that in the chain of commutator groups of the radical of G all non-compact commutator factor groups are purely incompact (i.e. contain no nontrivial compact subgroup) and appear last in the series.

Theorem X and its corollaries are special cases of Theorem IX.

Corollary VIII.1 and the following corollary will both be generalized in

Section XII (Corollary XII.5). The present statement uses only the known fact that a connected locally compact maximally almost periodic group is a direct product of a vector group and a compact connected group (for a precise definition of maximal almost periodicity see Section XII).

COROLLARY IX.2. *Let G be a locally compact or first countable topological group with a closed normal subgroup N isomorphic to a reflexive Banach space such that G/N is a connected locally compact maximally almost periodic group; then there is a compact subgroup C of G and a vector solvable normal subgroup S (solvable of length 2) such that G is the semidirect product of S and C, and all such groups C are conjugate.*

PROOF. Because of the characterization of maximally almost periodic groups mentioned above [23, p. 129] there are normal subgroups S and C_o of G intersecting in N such that S/N is a finite dimensional vector subgroup and C_o/N is compact. Therefore S is vector solvable of length 2 and G/S is isomorphic to C_o/N and is therefore compact. By Theorem IX, G is the semidirect product of S and a compact group (incidentally isomorphic to C_o/N), and all compact groups sharing this property are conjugate.

IX. A DIGRESSION INTO LOCALLY COMPACT GROUPS.

9.1. LEMMA. *Let G be a locally compact connected abelian Lie group and $\mathfrak{A}$ a group of continuous automorphisms compact in the compact open topology. Then there is a vector subgroup N and a characteristic compact subgroup C such that G is the direct product NC and N and C are invariant under $\mathfrak{A}$.*

PROOF. It is well known that G contains an n-dimensional vector subgroup N' and a compact subgroup C isomorphic to an m-dimensional torus such that G is the direct product of N' and C [22, p. 110]. Since C is characteristic, it is clearly invariant under $\mathfrak{A}$. We have to show that we can find N' invariant under $\mathfrak{A}$.

There is an open homomorphism f of the $n+m$-dimensional euclidean vector space R^{n+m} onto G which is a local isomorphism. Any continuous automorphism α of G can be lifted in an unique fashion to an automorphism α' of R^{n+m} (i.e. to a linear transformation) such that $f\circ\alpha' = \alpha\circ f$. In order to prove this assertion, let U be a neighborhood of 0 in R^{m+n} which is homeomorphically mapped into G by f. Let V be a neighborhood of 1 in G such that $V \cap \alpha V$ contains $f(U)$. Let α_o be the restriction of α to V. Then $\alpha'_o = f^{-1}\circ\alpha_o\circ f$ is a well-defined local automorphism of R^{n+m} which can be straightforwardly extended to an automorphism α' of R^{m+n} by linearity. Let r be in R^{n+m}. Then there exists a natural number k such that $(1/k)r$ is in U. Then we have $\alpha(f(r)) = k\alpha(f((1/k)r)) = kf(\alpha'((1/k)r)) = f(\alpha'r)$. This proves the assertion.

The mapping $\alpha \to \alpha'$ is an isomorphism algebraically and topologically of the group of all automorphisms of G into the group of all automorphisms of R^{n+m}. The algebraic part of this assertion is clear from the construction and the topological part follows from the observation that an automorphism α is close to the identity automorphism with respect to the topology of uniform convergence on compact sets if and only if α' is close to the identity of R^{n+m} relatively to uniform convergence on compact sets.

Therefore α is mapped by $\alpha \to \alpha'$ onto a compact group of automorphisms of R^{m+n} leaving the subspace $f^{-1}(C)$ invariant, since C is invariant under $\mathfrak{A}$. By 7.2. there is an inner product on R^{n+m} such that $\mathfrak{A}$ is a compact group of orthogonal transformations. Since they leave the subspace $f^{-1}(C)$ invariant, they do so with an orthogonal complement R_n. Since the kernel of f is entirely contained in $f^{-1}(C)$, the subspace R_n is mapped isomorphically onto a vector subgroup N of G which, because of the identity $\alpha \circ f = f \circ \alpha'$, is invariant.

9.2. LEMMA. *Let G be a compact group and $\mathfrak{A}$ a group of continuous automorphisms compact in the compact open topology (or, in this case equivalently, in the uniform topology). Then G is a projective limit of factor groups G/H, which are Lie groups and have the property that H is invariant under $\mathfrak{A}$, so that there is a natural action of $\mathfrak{A}$ on G/H.*

PROOF. Let $G \otimes \mathfrak{A}$ be the holomorph of G with $\mathfrak{A}$, i.e. G is the cartesian product of G and $\mathfrak{A}$ with the following multiplication:

$$(g,f)(g',f') = (g^{f'}g', ff').$$

It is well known and easily checked that $G \times \mathfrak{A}$ with this multiplication is a compact topological group, having a normal subgroup $G \times \{1\}$ isomorphic to G, whose factor group $(G \otimes \mathfrak{A})/(G \times \{1\})$ is isomorphic to $\mathfrak{A}$, and the elements of $1 \times \mathfrak{A}$ act as inner automorphisms on $G \times \{1\}$. The compact group $G \otimes \mathfrak{A}$ is a projective limit of Lie groups $(G \otimes \mathfrak{A})/K$. Then the groups H such that $H \times \{1\} = (G \times \{1\}) \cap K$ are normal in $G \otimes \mathfrak{A}$, and G/H is isomorphic to $(G \times \{1\})K/K$, which is a closed subgroup of $(G \otimes \mathfrak{A})/K$ and is therefore a Lie group. Thus G is the projective limit of groups G/H which satisfy the requirements of the lemma.

9.3. **Lemma.** *Let G be a locally compact abelian group which is the direct product of an n-dimensional vector group N and a compact group C. Suppose that $\{E_i : i \in I\}$ is a descending chain of closed subgroups such that $E_i/(E_i \cap C)$ is an n-dimensional vector group. Let $E = \cap\{E_i : i \in I\}$. Then $E/(E \cap C)$ is an n-dimensional vector group.*

Proof. C is the unique maximal compact group. On factoring by $E \cap C$ we may assume that E contains no compact subgroups. The image of E in G/C is then an isomorphic image; therefore E is a subgroup of an n-dimensional vector group. Its connected component E' is then an m-dimensional vector group, $m \leq n$, and we have to prove $m = n$. On factoring by E' and calling the quotient G again, we may now assume that E is discrete. Then we have to show that $n = 0$.

Let V be a compact neighborhood of 1 in G such that $V^2 \cap E = \{1\}$. Then the intersection of the collection of compact subspaces $V^2 \cap E_i$ is $\{1\}$. Hence there is an index $j \in I$ such that $V^2 \cap E_j \subset V^2$, which implies $E_j = (E_j \cap V^2) \cup (E_j \cap (G \setminus V^2))$. Since $E_j/(E_j \cap C)$ is isomorphic to an n-dimensional vector group, there is an n-dimensional vector subgroup N_j in E_j such that $E_j = N_j(E_j \cap C)$ [22, p. 110]. Using the previous decomposition of E_j in a disjoint sum of open sets we get $N_j = (N_j \cap V^2) \cup (N_j \cap (G \setminus V^2))$. Since N_j is connected, one of these open sets has to be empty, but the first contains 1. Hence $N_j \subset V$ and is therefore compact. Hence the dimension n of the vector group N_j is zero.

Theorem X. *Let G be a locally compact connected abelian group and $\mathfrak{A}$ a compact group of automorphisms of G (w.r.t. the compact open topology). Then G contains a vector subgroup N and a compact subgroup C such that G is the direct product of N and C and both N and C are invariant under $\mathfrak{A}$.*

Proof. G contains a maximal connected compact characteristic subgroup C which is obviously invariant, and it contains a vector group N' such that G is the direct product of N' and C. We have to show that N' can be found invariant.

By 9.3., C is a projective limit of Lie groups C/H such that H is invariant under $\mathfrak{A}$. Then $G/H = N'C/H$ is a Lie group on which $\mathfrak{A}$ acts in a natural fashion so that Lemma 9.1. applies. Hence there exists a subgroup E_H of G containing H such that E_H/H is a vector group isomorphic to N' which is invariant under the action of $\mathfrak{A}$ on G/H; hence E_H itself is invariant. Furthermore, G/H is the direct product of E_H/H and C/H. Thus $E_H \cap C$ is H.

Let $\mathfrak{g}$ be the collection of closed subgroups E of G having the following properties:

(i) E is invariant under $\mathfrak{A}$.

(ii) $E/(E\cap C)$ is isomorphic to the vector group N' which in turn is isomorphic to G/C.

We show that $\mathfrak{g}$ is inductive when partially ordered by inclusion. Let $\{E_i : i \in I\}$ be a totally ordered descending chain. Its intersection E is closed and invariant under $\mathfrak{A}$ since all E_i are closed and invariant. Thus, clearly, $E\cap C$ is a compact invariant group and by Lemma 9.3., E satisfies (ii).

Since $\mathfrak{g}$ is inductive, there is a minimal element N. All we have to show in completing the proof is that $N\cap C=\{1\}$. But $N\cap C$ contains arbitrarily small invariant subgroups H by 9.2. such that $N=N_H(N\cap C)$ with an invariant group N_H intersecting C in H (see above); since N_H/H is isomorphic to G/C we know that N_H belongs to $\mathfrak{g}$ and hence by minimality of N is equal to N, which implies that $N_H\cap C=N\cap C=H$ for arbitrarily small subgroups H. Thus we get indeed $N\cap C=\{1\}$.

COROLLARY X.1. *Let G be a locally compact group which is a direct product of a vector subgroup N_1 and a compact group C. Let $\mathfrak{A}$ be a compact group of automorphisms of G (w.r.t. the compact open topology). Then G contains a vector subgroup N such that G is the direct product of N and C, and N and C are invariant under A.*

PROOF. Since C is in this case the unique maximal compact subgroup, it is characteristic and therefore invariant. Let G_1 be the connected component of 1 in the center of G; then G_1 contains N' and is a characteristic closed subgroup, which is therefore invariant under $\mathfrak{A}$. Hence by Theorem X it contains a vector subgroup N invariant under $\mathfrak{A}$ which is isomorphic to $G_1/(G_1\cap C)$ and hence to $G_1 C/C=G/C$. Therefore $G=NC$ and this product is direct since N is in the center of G.

COROLLARY X.2. *Let G be a locally compact group and F a closed normal subgroup such that G/F is compact. Let F_1 be the component of the center of F. Then F_1 is the direct product of a vector subgroup N and a compact connected abelian group C and both N and C are normal in G.*

PROOF. As a locally compact abelian group, F_1 is isomorphic to the direct product of a vector subgroup N_1 and a compact connected abelian group C. The centralizer of F_1 contains F. Hence, by Lemma 9.1., G/F acts as a compact group of automorphisms on F_1, and a subgroup of F_1 is normal in G if and only if it is invariant under G/H by this action. By Theorem X, F_1 is the direct product of a vector group N and a compact group C both of which are invariant under G/H. This proves the corollary.

9.4. PROPOSITION. *Let G be a locally compact group. If G is the semidirect product of a compact normal group C and a vector group N_1, then G is the direct product of C and a vector subgroup N isomorphic to N_1.*

PROOF. (a) We shall show first that all automorphisms of C induced by inner automorphisms with elements from N_1 are induced by inner automorphisms with elements from C. Let $\mathfrak{A}$ be the full automorphism group of C and $\mathfrak{F}$ the normal subgroup of inner automorphisms; let furthermore $\mathfrak{G}$ be the group of automorphisms of C induced by inner automorphisms of G. Then $\mathfrak{A}/\mathfrak{F}$ is totally disconnected [9] and with it is its subgroup $\mathfrak{G}/\mathfrak{F}$. If Z is the centralizer of C in G, then $\mathfrak{G}$ is isomorphic to G/Z, and $\mathfrak{F}$ is isomorphic to CZ/Z (or, equivalently, to the quotient of C modulo its center $C \cap Z$). Hence G/CZ is isomorphic to $\mathfrak{G}/\mathfrak{F}$ on one hand, and is, therefore, totally disconnected; it is, on the other hand, a homomorphic image of G/C which is isomorphic to N_1 and is, therefore, connected. Hence, $\mathfrak{G}/\mathfrak{F}$ must contain only one point, i.e. $\mathfrak{G} = \mathfrak{F}$ and $G = CZ = ZC$.

(b) Let ϕ be the representation of N into $\mathfrak{g}$, which assigns to each $n \in N$ the automorphism $c \rightarrow n^{-1}cn$ of C. From (a) we know that ϕ maps N actually into $\mathfrak{F}$, but $\mathfrak{F}$, being isomorphic to $C/(C \cap Z)$, is compact. Thus G satisfies the hypotheses of a lemma about maximal almost periodic groups which the reader will find in the section about maximal almost periodic groups as Proposition 12.14; it may be observed that its proof is independent from other developments in that section. By this proposition, G is maximally almost periodic. Then the component G_o of G is also maximally almost periodic and is, by a theorem of Freudenthal [6; 22] a direct product of a compact group and a vector group N_2. Since G_o contains N_1, the quotient G/G_o is compact. By Corollary X.2 we can represent the center of G_o as a direct product of a vector group N and some compact group, both of which are normal in G. Since N does not contain nontrivial compact subgroups, $N \cap C = \{1\}$; hence the product of the two normal subgroups N and C is direct. The intersection of C with the component G_o is the characteristic subgroup C_o. Thus G_o is entirely contained in CN and so is N_1. Hence $CN' = G$ is contained in CN. This proves $G = NC$. Both N and N_1 are isomorphic to the quotient G/C; thus they have to be isomorphic. This completes the proof.

COROLLARY X.3. *Let G be a locally compact group and F a closed normal* *subgroup which is the semidirect product of a normal compact subgroup C_o and a vector group N_o. Suppose that G/F is compact. Then G is the semidirect product of a normal vector subgroup N (isomorphic to N_o) and a compact group C containing C_o. If D is another compact group such that $G = ND$, $C_o \subset D$, then C and D are conjugate.*

If, in addition, G/F is totally disconnected (i.e. F contains the component of G), then C contains an open subgroup C_1 of finite index normal in G such that NC_1 is a direct product (and, clearly, NC_1 is of finite index in G).

PROOF. By Proposition 9.4., we may assume that F is the direct product of N_o and C_o. Then, using Corollary X.2., we know that F contains a vector subgroup N normal in G and isomorphic to N_o. Then, by Corollary VIII.1, G is the semidirect product of N and some compact subgroup C. Because C_o is the unique maximal compact subgroup of F, the intersection of C with F is indeed C_o. Suppose now that G/F is totally disconnected. Let Z be the centralizer of N in G which necessarily contains C_o. Let $C_1 = Z \cap C$. Then C/C_1 is isomorphic to $CZ/Z = G/Z$ which acts as a compact effective group of automorphisms on the vector group N (7.1.). Hence, by 7.2., C/C_1 is isomorphic to a closed subgroup of some orthogonal group. Since it is totally disconnected, it has to be discrete, and therefore finite. C_1 is normal in C since Z is normal in G; but C_1 is also the maximal compact group in Z hence normal in Z. So it is normal in G. This finishes the proof, since the conjugacy of C and D is a direct consequence of Theorem IV.

COROLLARY X.4. *Let G be the semidirect product of a compact normal subgroup C and a connected closed subgroup N_1. Then either of the following conditions is sufficient that there is a normal subgroup N isomorphic to N_1 such that G is the direct product of C and N:*
   (i) N_1 *is a vector group,*
   (ii) *the center of C is trivial.*

PROOF. The sufficiency of condition (i) was established in 9.4. We show that (ii) is sufficient. Let Z be the centralizer of C in G. After the proof of 9.4. we know $G = ZC$. But $Z \cap C$ is the center of C, hence $Z \cap C = \{1\}$, and $N = Z$ satisfies the requirement.

X. SPLITTING CENTRAL VECTOR GROUPS IN TOPOLOGICAL GROUPS.

10.1. LEMMA. *Let G be a topological group (which we write additively although it need not be commutative), and let N be a closed central subgroup such that*

G/N is abelian. Then the mapping $(g,h) \to [g,h] = g+h-g-h$ is a continuous skew symmetric bilinear mapping of G into N (i.e. $[g+g',h] = [g,h] + [g',h]$ and $[g,h+h'] = [g,h] + [g,h']$); its values depend only on the cosets of g and h modulo N. Hence the mapping defined by $\{x,y\} = [g,h]$, where x,y are cosets modulo N and g and h arbitrary elements of x and y respectively, is a continuous bilinear mapping of G/N into N. The closure of the subgroup generated in N by $\{G/N, G/N\}$ is the commutator subgroup G' of G.

PROOF. In order to show the bilinearity of $(g,h) \to [g,h]$, it is sufficient to show linearity in one argument, since $[g,h] = -[h,g]$. From the definition we have $h^{-g} = [g,h] + h$. Since G/N is abelian, addition is commutative modulo N, which shows that $[g,h]$ is in N. We compute $(h+h')^{-g}$ in two different ways:

$$(h+h')^{-g} = [g,h+h'] + (h+h') \text{ and}$$

$$(h+h')^{-g} = h^{-g} + h'^{-g} = [g,h] + h + [g,h'] + h'$$

$$= ([g,h] + [g,h]) + (h+h')$$

since N is central. Equating yields the bilinearity in the second argument.

If n and n' are elements in N, then indeed $[g+n, h+n'] = [g,h]$ because N is central. This shows that $(x,y) \to \{x,y\}$ is a well-defined mapping, and it is continuous since $(g,h) \to [g,h]$ is continuous.

All commutators are of the form $\{x,y\}$ with some x,y in G/N which shows that the commutator subgroup G' of G is indeed the closure of the subgroup generated by $\{G/N, G/N\}$.

10.2. LEMMA. *Let the hypotheses be the same as before and suppose in addition that G/N is an n-dimensional vector space and that N is a topological vector space. Then G' is an abelian Lie group of at most dimension $(n;2)$.*

PROOF. It is easy to see with the current methods that the mapping $(x,y) \to \{x,y\}$ being bilinear over the ring of integers is bilinear over the rationals, i.e. that $\{n_x/m, y\} = (n/m)\{x,y\}$, for this identity is equivalent to $m\{n_x/m, y\} = n\{x,y\}$. By continuous extension $(x,y) \to \{x,y\}$ is bilinear over the reals. It is therefore decomposable into a bilinear mapping of G/N to the exterior product $G/N \wedge G/N$ over the reals, and a linear mapping from $G/N \wedge G/N$ into N. But the exterior product $G/N \wedge G/N$ over the reals is $(n;2)$-dimensional. This proves the statement since a finite dimensional subspace is closed in any topological vector space containing it.

10.3. **LEMMA.** *If G' is a finite dimensional vector subspace of a locally convex topological vector space N over the reals, then there is a subspace N_o such that N is the direct vector space sum of N_o and G'.*

PROOF. This follows by finite induction from the fact that any one dimensional subspace admits a supplement by the Hahn Banach theorem.

10.4. **PROPOSITION.** *Let G be an abelian topological group and N a closed subgroup isomorphic to the additive group of a reflexive Banach space such that G/N is a finite dimensional vector group. Suppose that N admits a cross section on G. Then G contains a finite dimensional vector group F and G is the direct product of N and F. (We write groups multiplicatively again.)*

PROOF. The proposition follows by induction on the dimension of G/N from the validity of the lemma if G/N is one dimensional; this we shall prove now: G/N contains an infinite cyclic subgroup modulo which G/N is compact. Let Z be its full counter image in G. Let z be an element in the coset modulo N in Z which generates the infinite cyclic group Z/N. Then z generates an infinite cyclic group D in G and $Z = ND$. The subgroup ND/D admits a cross section in G/D, and $(G/D)/(ND/D)$ is isomorphic to G/Z and to $(G/N)/(Z/N)$ and is therefore isomorphic to the compact circle group. By Theorem VII, G/D contains a subgroup F_1 isomorphic to the circle group and such that G/D is the direct product of F_1 and ND/D. Let F be the full counter image of F_1 in G. Then F, being a one-dimensional subgroup of a vector subgroup, is isomorphic to the reals since D is cyclic, and $G = NDF = NF$, since $ND \cap F = D$, we have indeed $N \cap F = \{1\}$.

10.5. **PROPOSITION.** *Let G be a topological group, N a closed central subgroup of G satisfying the following conditions:*
  (i) *N is isomorphic to the additive group of a reflexive Banach space,*
  (ii) *G/N is an n-dimensional vector group,*
  (iii) *N admits a cross section in G.*
Then G contains an at most $n(n+1)/2$-dimensional nilpotent Lie group L, nilpotent of class 2 and homeomorphic to a euclidean space, N contains a subgroup N_o isomorphic to the additive group of a reflexive Banach space such that
  (i') *G is the direct product of N_o and L,*
  (ii') *$N \cap L$ is central in L, $L/(N \cap L)$ is isomorphic to G/N and $N \cap L$ is the commutator group of G.*

PROOF. By 10.1. and 10.2., the commutator group is an at most $(n;2)$-

dimensional vector subspace G' of N. By 10.3. there is a subgroup N_o of N isomorphic to the additive group of a reflexive Banach space such that N is the direct product of N_o and G'. The group G/G' satisfies the hypotheses of Proposition 10.4. with N/G' instead of N. Hence there is a normal subgroup L in G containing G' such that G/G' is the direct product of N/G' and L/G', and $L \cap N$ is equal to G'; furthermore L/G' is isomorphic to $(G/G')/(N/G')$ which in turn is isomorphic to G/N. Thus L is a Lie group, its dimension is the sum of the dimensions of $G' = N \cap L$ and of G/N which is n.   Clearly L is nilpotent of class 2, since modulo its center, which contains G', it is abelian. Obviously $NL = N_o L = G$ and $N_o \cap L = \{1\}$. L is homeomorphic to $(N \cap L) \times G/N$ since $N \cap L$ admits a cross section in L (7.5.).

10.6. **Proposition.** *Let G be a topological group, N a closed central subgroup satisfying*

  (i) *N is isomorphic to a reflexive Banach space,*

  (ii) *G/N is compact,*

  (iii) *N admits a cross section in G.*

Then G is isomorphic to the direct product of N and G/N.

Proof. By Theorem VII there is a subgroup B of G isomorphic to G/N such that G is the semidirect product of N and C; but since N is central this product is direct.

Theorem XI. *Let G be a topological group, N a closed central subgroup and let the following conditions be satisfied:*

  (i) *N is isomorphic to the additive group of a reflexive Banach space,*

  (ii) *G/N is locally compact and the product of an n-dimensional vector group and a compact group,*

  (iii) *N admits a cross section in G.*

Then N contains a closed subgroup N_o isomorphic to a reflexive Banach space, G contains a nilpotent Lie group L, nilpotent of class 2 and of dimension less than or equal to $n(n+1)/2$ and homeomorphic to a euclidean space and a compact group C isomorphic to the compact group in (ii) such that $G = N_o L C$ and this product is topologically direct (however not algebraically in general). Further $L/(N \cap L)$ is a vector group of n dimensions, and $N_o L$ is a direct product algebraically as well as topologically.

Proof. Let $G/N = F/N \cdot C_1/N$ with subgroups F and C_1 of G containing N such that F/N is a finite dimensional vector space and C_1/N a compact group. Then, by Proposition 10.5., $F = N_o L$ and this product is direct, L and

N_o having the properties indicated in the theorem. By Theorem VII, $C_1 =$ NC and this product is semidirect. Clearly $G = N_o LNC = N_o LC$ since N is normal and is contained in $N_o L$. Since N and L do not contain compact subgroups, $NL \cap C = \{1\}$.

COROLLARY XI.1. *Let G be a topological group with a closed central subgroup N satisfying the following two conditions:*

(i) *N is isomorphic to the additive group of a reflexive Banach space,*

(ii) *G/N is a locally compact group which is the direct product of a vector group of dimension n and a compact group,*

and any one of the following three conditions:

(iii)　1) *N admits a cross section in G,*

(iii)　2) *G is locally compact,*

(iii)　3) *G satisfies the first axiom of countability.*

Then G is algebraically and topologically a direct product of the following subgroups:

(a) *a closed subgroup N_o having the same properties as N,*

(b) *a nilpotent group L of dimension not exceeding $n(n+1)/2$,*

(c) *a compact group C isomorphic to the compact component of G/N.*

Moreover, $L/(N \cap L)$ is an n-dimensional vector group.

PROOF. The groups N_o, L, and C exist by Theorem XI. We have only to show that the elements of L and C commute. The mapping $(f,c) \to [f,c]$ is bilinear from $F \times C$ into N. This is proved as in Lemma 2.1 since the elements of F and C commute modulo N and N is central. Hence, for all $f \in F$, the compact set $[f,C]$ is a compact subgroup of N which has to be trivial since N does not contain nontrivial compact subgroups. Hence $[f,c] = 1$ for all $f \in F$ and $c \in C$, which proves the corollary.

COROLLARY XI.2. *Let G be a topological group with a closed normal subgroup N. Let Z be its centralizer in G. Suppose the following conditions are satisfied:*

(i) *N is isomorphic to the additive group of a reflexive Banach space,*

(ii) *Z/N is a semidirect product of a compact normal group C_1' and a vector group of n dimensions, and G/Z is a compact group,*

(iii) (1) *or* (iii) (2) *or* (iii) (3) *as in Corollary XI.1.*

Then G contains the following groups:

(a) *a closed subgroup N_o of N having the same properties as N,*

(b) *a nilpotent Lie group L of dimension at most $n(n+1)/2$ which is homeomorphic to a euclidean space,*

(c) *a compact group C,*

and $G = N_oLC$; this product is topologically direct, N_oL is a direct product and $(N_oL)C$ is a semidirect product. Moreover, $N/(N\cap L)$ is an n-dimensional vector space, and C contains a subgroup C_1 such that $Z = N_oLC_1$ is a direct product and C/C_1 is isomorphic to G/Z.

PROOF. Clearly condition (ii) implies that G/N is a locally compact group. It satisfies the hypotheses of Corollary X.3. Hence, G/N is the semidirect product of a vector group F'/N and a compact group C'/N. By Proposition 10.5. $F' = N_oL$ and by Theorem VII, $C' = NC$, so that $G = N_oLC$ as asserted. The compact group C'/N contains C_1'; let C_1 in C be such that $Z = N_oLC_1$; then this product is direct by Corollary XI.1. The remainder of the corollary is obvious.

XI. EXAMPLES OF NON-SPLITTING EXTENSIONS.

We have already seen holomorphic, or splitting, extensions (cf. the proof of Lemma 9.2.). In this section we give examples of non-splitting extensions. They will be used to illustrate the limitations on the degree of generality that one may expect in questions of the type considered in this paper.

11.1. EXAMPLE. Let Q be a topological ring and A and B two topological Q-modules (written additively). Let β be a continuous bilinear mapping of the Q-module $B\times B$ into the Q-module A. On the cartesian product $A\times B$ we define the following multiplication:

$$(a,b)(a',b') = (a+a'+\beta(b,b'),b+b').$$

This multiplication turns the product space $A\times B$ into a topological group which we shall denote by $G(A,B,\beta)$. Furthermore, $G(A,0,\beta)$ is a submodule of $G(A,B,\beta)$ isomorphic to A and is normal. Moreover, the quotient

$$G(A,B,\beta)/G(A,0,\beta)$$

is isomorphic to B.

PROOF. (a) First we show that the multiplication is associative:

$$((a,b)(a',b'))(a'',b'') = (a+a'+\beta(b,b'),b+b')(a'',b'')$$

$$= (a+a'+a''+\beta(b,b')+\beta(b+b',b''),b+b'+b'');$$

on the other hand we get

$$(a,b)((a',b')(a'',b'')) = (a,b)(a'+a''+\beta(b',b''),b'+b'')$$

$$= (a+a'+a''+\beta(b',b'')+\beta(b,b'+b''),b+b'+b'')$$

which coincides because of the bilinearity of β, indeed with the preceding expression.

(b) If $(a,b)(a',b') = (a+a'+\beta(b,b'), b+b') = (0,0)$, then $b' = -b$ and $a' = -a+\beta(b,b)$; i.e. the inverse exists and is unique.

(c) Since A and B are topological modules and β is continuous, both the multiplication and the inversion are continuous functions. This finishes the proof that $G(A,B,\beta)$ is a topological group.

(d) That $G(A,0,\beta)$ is isomorphic to A under $a\to(a,0)$ is trivial.

(e) Clearly, if either b or b' is zero, since β is bilinear, $(b,b') = 0$. This, then, immediately shows that $G(A,0,\beta)$ is normal.

(f) That $G(A,B,\beta)/G(A,0,\beta)$ is isomorphic to B is also immediate.

11.2. EXAMPLE. Let A,B,β be as before. Let $g = (a,b)$ and $g' = (a',b')$ be in $G(A,B,\beta)$ and let $[g,g']$ be the commutator of g, and g' in $G(A,B,\beta)$. Then

$$[g,g'] = (\beta(b,b')-\beta(b',b),0)$$

and

$$g'gg'^{-1} = (a+\beta(b',b)-\beta(b,b'),b).$$

PROOF. We observe that

$$gg' = (a+a'+\beta(b,b'), b+b') \quad \text{and} \quad g'g = (a+a'+\beta(b',b), b+b').$$

Furthermore,

$$(g'g)^{-1} = (-a-a'-\beta(b',b)+\beta(b+b',b+b'), -b-b').$$

Thus

$$[g,g'] = (gg')(g'g)^{-1} = (a+a'+\beta(b,b')-a-a'-\beta(b',b)+$$
$$\beta(b+b',b+b')+\beta(b+b',-b-b'),0) = (\beta(b,b')-\beta(b',b), 0).$$

This proves the first statement. Clearly $g'gg'^{-1} = [g',g]g$, thus $g'gg'^{-1} = (\beta(b,b')-\beta(b,b')+a+\beta(0,b), 0+b)$ which proves the second statement.

11.3. EXAMPLE. Let $G(A,B,\beta)$ be defined as before. Then the (topological) commutator group $K(A,B,\beta)$ is the closure of the submodule of $G(A,0,\beta)$ generated by all elements of the form $(\beta(b,b')-\beta(b',b),0)$. The center $Z(A,B,\beta)$ of $G(A,B,\beta)$ is the set of all elements (a,b) such that

$$\beta(b,b') - \beta(b',b) = 0.$$

In particular we have

(i) $\qquad\qquad\qquad K(A,B,\beta) \subset G(A,0,\beta),$

(ii) $\qquad\qquad\qquad G(A,0,\beta) \subset Z(A,B,\beta).$

If β is symmetric, then $G(A,B,\beta)$ is a module. Otherwise $G(A,B,\beta)$ is nilpotent of class 2.

PROOF. All statements except the last one are corollaries to 11.2; but the last one follows from the fact that $G(A,B,\beta)/G(A,0,\beta)$ which is isomorphic to B, is abelian. Hence $G(A,B,\beta)/Z(A,B,\beta)$ is certainly abelian.

11.4. Let A,B and $G(A,B,\beta)$ be as before and suppose in addition that the module A is uniquely divisible by 2; i.e. for each element $a \in A$ there is an element a' such that $a'+a'=a$; we shall denote a' by $a/2$. Let us define on $A \times B$ the addition

$$g \oplus g' = (a,b) \oplus (a',b') = (a+a'+(1/2)(\beta(b,b')+\beta(b',b)), b+b').$$

The algebraic structure defined on $A \times B$ by this addition and the multiplication $[g,g']=$ commutator of g and g' in $G(A,B,\beta)$ will be denoted by $\mathfrak{g}(A,B,\beta)$. Then $\mathfrak{g}(A,B,\beta)$ is a Lie ring and

(i) $gg' = g \oplus g' \oplus (1/2)[g,g']$ (Cambell-Hausdorff formula),

(ii) $[g,g']=gg'g^{-1}g'^{-1}$ and $g \oplus g' = gg'((1/2)[g',g])$.

PROOF. The mapping $(b,b') \to (1/2)(\beta(b,b')+\beta(b',b))$ is a symmetric bilinear mapping from $B \times B$ into A. Therefore, by 11.1. and 11.3., the addition turns $A \times B$ into a topological Q-module.

It is immediately clear that $g \oplus g^{-1} = (0,0)$, so that the inverses of the addition $\oplus$ and the multiplication in $G(A,B,\beta)$ are the same. For elements in $G(A,0,\beta)$ we have $g^{-1} = -g$, so that $[g,g']= -[g',g]=[g',g]^{-1}$. For all g,g', $h \in \mathfrak{g}(A,B,\beta)$, clearly $[[g,g'],h]= (0,0)$; this shows that the Jacobi identity is trivially satisfied.

We prove distributivity after we have established the identities (i) and (ii):

(i) $\qquad (a,b) \oplus (a',b') \oplus (1/2)[(a,b),(a',b')]$

$$= (a+a'+(1/2)(\beta(b,b')+\beta(b'b))+(1/2)(\beta(b,b')-\beta(b',b)), b+b')$$

$$= (a,b)(a',b').$$

(ii) $(a,b)(a',b') \cdot (1/2)[(a',b'),(a,b)]$

$$= (a+a'+\beta(b,b'),b+b')((1/2)\beta(b',b) - (1/2)\beta(b,b'),0)$$

$$= (a+a'+(1/2)(\beta(b,b')+\beta(b',b)),b+b') = (a,b) \oplus (a',b').$$

Let now g, g' and h be elements in $\mathfrak{g}(A,B,\beta)$. Then

$$[(g \oplus g'),h] = [gg'(1/2)[g',g],h] = [gg',h](1/2)[[g',g],h]$$

because of Lemma 10.1. The second commutator is zero, and, again by 10.1.

$$[gg',h] = [g,h][g',h].$$

On the other hand $[g,h][g',h'] = [g,h] \oplus [g',h]$ since the second coordinate of every commutator is zero. This proves distributivity and finishes the argument.

11.5. **EXAMPLE.** Let H be a subset of $A \times B$ such that $H \cap (A \times \{0\})$ is divisible by 2. Then H is a subgroup of $G(A,B,\beta)$ if and only if it is a subring of $\mathfrak{g}(A,B,\beta)$. It is a normal subgroup in $G(A,B,\beta)$ if and only if it is an ideal in $\mathfrak{g}(A,B,\beta)$.

PROOF. The first half of the statement follows immediately from 11.4. (i), (ii). The second follows from

$$g'gg'^{-1} = [g,g']g = [g,g'] \oplus g.$$

11.6. **EXAMPLE.** If β is a skew symmetric bilinear form, i.e., if $\beta(b,b') = -\beta(b',b)$, then $\mathfrak{g}(A,B,\beta)$ is a Lie algebra over Q, and the addition and the Lie ring addition coincide with the addition on the product module $A \times B$. Moreover, $[(a,b), (a',b')] = (2\beta(b,b'), 0)$.

PROOF. By the definition of $\oplus$ in 11.4. we get immediately

$$(a,b) \oplus (a',b) = (a+a', b+b').$$

From this observation it becomes clear that $\mathfrak{g}(A,B,\beta)$ is a Lie algebra over Q.

11.7. **EXAMPLE.** Let $A,B,G(A,B,\beta)$ be as in 11.4. Suppose there is a subgroup H such that

$$G(A,B,\beta) = G(A,0,\beta)H, \quad G(A,0,\beta) \cap H = (0,0)$$

(i.e. $G(A,0,\beta)$ splits in $G(A,B,\beta)$). Then $G(A,B,\beta)$ is abelian and $\mathfrak{g}(A,B,\beta)$ is a Q-module. This occurs if and only if β is symmetric.

PROOF. $\mathfrak{g}(A,B,\beta)$ is the direct sum of the ideal $\mathfrak{g}(A,0,\beta)$ and the subalgebra $\mathfrak{H}$, where $\mathfrak{g}(A,0,\beta)$ is the set $A \times \{0\}$ and $\mathfrak{H}$ the set H considered as subalgebras of $\mathfrak{g}(A,B,\beta)$. Since $\mathfrak{H}$ is isomorphic to $\mathfrak{g}(A,B,\beta)/\mathfrak{g}(A,0,\beta)$, it is abelian. $\mathfrak{g}(A,0,\beta)$ is abelian. But the ideal $[\mathfrak{g}(A,0,\beta),\mathfrak{H}\,]$ generated by all $[g,h]$ with $g \in \mathfrak{g}(A,0,\beta)$ and $h \in \mathfrak{H}$ is zero since $\mathfrak{g}(A,0,\beta)$ is in the center of $\mathfrak{g}(A,B,\beta)$. Hence $[\mathfrak{g}(A,B,\beta),\ \mathfrak{g}(A,B,\beta)\,]=(0)$. This shows that $\mathfrak{g}(A,B,\beta)$ and $G(A,B.\beta)$ are abelian. But commutativity means

$$(0,0)=[\,(a,b),(a',b')\,]=(\beta(b,b')-\beta(b',b),0)$$

for all a,a',b,b'. Hence, β is symmetric.

REMARK. Rank $B=1$ (as a Q-module) or char $Q=2$ are sufficient conditions for $G(A,B,\beta)$ to be abelian whatever A is if β is skew symmetric.

11.8. EXAMPLE. Let V_n be the n-dimensional real vector group (which we realize as a vector space product of n copies of the reals). Let β be the following bilinear form from V_2 into V_1:

$$\beta((b_1,b_2),\ (b_1',b_2'))=b_1b_2'-b_1'b_2.$$

Then $L=G(V_1,V_2,\beta)$ is a nilpotent Lie group of class 2 and of dimension 3, having the following properties:

(i) $G(V_1,0,\beta)$ is a one-dimensional vector subgroup which is the center and the commutator subgroup of L and does not split in G.

(ii) if W is any one-dimensional vector subgroup of V_2, then $G(V_1,W,\beta)$ is a two-dimensional vector subgroup.

(iii) L has no compact neighborhood invariant under all inner automorphisms.

PROOF. We have to verify the statement about the center and the commutator group. The commutator of two elements: $g,g' \in G(V_1,V_2,\beta)$ is $[g,g']=(2(b_1b_2'-b_1'b_2),\ 0)$. This proves (i). Statement (ii) follows from the fact that $G(V_1,\ 0,\ \beta)$ is the center of L. Statement (iii) follows from the observation that the commutator subgroup is not compact [9], [17]. (See also 12.4.).

This example illustrates the rôle of the nilpotent Lie group L in Theorem XI and its corollaries.

11.9. EXAMPLE. Let Z_n be a subgroup of V_n generated by the n independent elements $e_1,\cdots,e_n,\ \ e_i=(0,\cdots,0,1,0,\cdots,0)$ with 1 in the ith place. Let β_1 be the following bilinear form from V_2 into V_1/Z_1:

$$\beta_1((b_1,b_2),(b_1',b_2')) = (b_1 b_2' - b_1' b_2) + Z_1.$$

(In this case, Q is the integers.)

Then $K = G(V_1/Z_1, V_1, \beta_1)$ is isomorphic to $G(V_1, V_2, \beta_1)/G(Z_1, 0, \beta_1)$ and has the following properties:

(i) $G(V_1/Z_1, 0, \beta_1)$ is the center and the commutator of K and does not split in K,

(ii) if W is a one-dimensional vector subgroup of V_2, then $G(V_1/Z_1, W, \beta_1)$ is a cylinder group in K,

(iii) if U is any compact neighborhood of 0 in V_2, then $G(V_1/Z_1, U, \beta_1)$ is a compact neighborhood of the identity in K invariant under all inner automorphisms of K [see 17].

PROOF. Let π be the coset homomorphism of V_1 onto V_1/Z_1. Then $(x,y) \to (\pi(x),y)$ is a homomorphism of $G(V_1, V_2, \beta)$ onto $G(V_1/Z_1, V_2, \beta_1)$ with

$$G(Z_1, 0, \beta)$$

as kernel. Then (i) and (ii) follow immediately from (i) and (ii) of 11.5. By 11.2. the transform of an element $g = (x,y,z) \in G(V_1/Z_1, V_2, \beta_1)$ by an element $g' = (x',y',z')$ is $g'gg'^{-1} = (x+2(yz'-y'z),y,z)$. This implies that all sets containing $G(V_1/Z_1, 0, \beta_1)$ are invariant, hence (iii).

REMARK. This group is similar to the group given by G. Birkhoff as a Lie group having no faithful representation as a matrix group. (Bull. Amer. Math. Soc. **62**(1936), 883–888. See [15, p.191–192].)

11.10. EXAMPLE. Let V_n and Z_n be as in 11.8. and 11.9. Let β_2 be the following bilinear form from Z_2 into V_1:

$$\beta_2((b_1,b_2),(b_1',b_2')) = b_1 b_2' - b_1' b_2.$$

Then $M = G(V_1, Z_2, \beta_2)$ is a one-dimensional Lie group having the following properties:

(i) $G_o = G(V_1, 0, \beta_2)$ is the connected component of M, is the center of M, and does not split in M,

(ii) the commutator group of M is $G(Z_1, 0, \beta_2)$, which is isomorphic to the group of all integers and is, therefore, not compact.

(iii) if Z is any cyclic subgroup of Z_2, then $G(V_1, Z, \beta_2)$ is isomorphic to $V_1 \times Z$.

(iv) M has arbitrarily small invariant neighborhoods of the identity.

(v) for any natural number n let nZ_m be the subgroup of Z_m consisting of all n-fold multiples of elements of Z_m. Then $G(nZ_1,nZ_2,\beta_2)$ is a normal subgroup of M for all n and the factor group of M module this group is compact. Therefore, for any $g \in M$ there exists a homomorphism onto a compact group which does not map g on the identity.

PROOF. (i)–(iii) are easily verified similarly to the previous examples. Since G_o is an open central subgroup, all subsets of it are invariant, hence (iv). The transform of an element $g = (nx,ny,nz)$ with integers x,y,z by an element $g' = (x',y',z')$ is $g'gg'^{-1} = (nx + 2n(yz' + y'z),ny,nz)$ which shows that $G(nZ_1,nZ_2,\beta_2)$ is indeed normal. The factor group

$$G(V_1,Z_2,\beta_2)/G(nZ_1,nZ_2,\beta_2)$$

has a finite quotient modulo the compact normal subgroup

$$G(V_1,nZ_2,\beta_2)/G(nZ_1,nZ_2,\beta_2)$$

and is, therefore, compact. For any (x,y,z) in M, there is always a natural number n such that $(x,y,z) \notin G(nZ_1,nZ_2,\beta_2)$.

REMARK. In a later section we shall particularly refer to the fact that M has small invariant neighborhoods and is maximally almost periodic. All the examples L,K,M are generated by a compact neighborhood of the identity. Clearly, M is isomorphic to a subgroup of L.

The following example is a generalisation of 11.9. and of an example given by Iwasawa [9, p. 550].

11.11. EXAMPLE. Let A be any compact connected abelian group with weight not exceeding the cardinality c of the continuum. (The weight of a topological space is the smallest cardinal which is a cardinal of a base for its topology.) Let f be a continuous one-to-one representation of V_1 onto a dense subgroup of A. Let β be defined as follows:

$$\beta((b_1,\cdots,b_{2n}), (b_1',\cdots,b_{2n})) =$$

$$f\left(\sum_{i=1}^{n} (b_{2i-1}b_{2i}' - b_{2i-1}'b_{2i}) \right).$$

Then $G(A,V_{2n},\beta)$ is a locally compact nilpotent group of class 2 with $G(A,0,\beta)$ as center and commutator group. (The algebraic commutator group is the dense subgroup $G(f(V_{2n}),0,\beta)$ of the commutator group.)

PROOF. It is only necessary to refer to the existence of the representation f under the given hypotheses. (See [19, p. 29]. The necessary changes are straightforword.)

XII. LOCALLY COMPACT GROUPS WITH INVARIANT NEIGHBORHOODS OF THE IDENTITY.

12.1. DEFINITION. A locally compact group G is said to have an *invariant compact neighborhood* if
  (A)   there is a compact neighborhood of the identity which is mapped onto itself by all inner automorphisms of G.
It is said to have *small* invariant neighborhoods if
  (B)   there is a basis of compact neighborhoods of the identity each of which is mapped onto itself by all inner automorphisms of G.
The group G is called *maximally almost periodic,* if
  (C)   the ring of all almost periodic continuous functions of G to the reals separates points.
The group G is said to be *representable into a compact group* if
  (D)   there is a one-to-one representation of G onto a dense subgroup of a compact group G^*. (This is equivalent to the statement that there are sufficiently many unitary representations to separate points.)
The following proposition gives some of the relations between the notions introduced in 12.1.

12.2. PROPOSITION.
  (i) (B) *implies* (A).
  (ii) *If* (A), *then there is a compact normal subgroup such that the quotient group satisfies* (B).
  (iii) (C) *is equivalent to* (D).
  (iv) *If G is generated by a compact neighborhood of the identity, then* (C) *implies* (B).
  (v) (A) *and* (C) *imply* (B).

REMARK. In one of the corollaries to the main theorem of this section we shall show that, provided the quotient of G modulo its component is compact, (B) implies (C).
  PROOF. (i) is trivial.
  (ii) This has been proved by Iwasawa and Yamabe, see [10].
  (iii) For this classic result we refer to [23].
  (iv) The validity of this assertion has been stated by Weil, [23, p. 129],

who gave an indication of a proof. We believe, however, that the proof
needs some additional details, and we give it, therefore, in the following:

Let W be a compact neighborhood generating G and let $f:G \to G^*$ be a
one-to-one representation of G into the compact group G^*. As a compact
group, G^* has the property (B); hence there is a basis $\mathfrak{U}$ of compact invariant
neighborhoods of the identity in G^*. For each V let U_V be $f^{-1}(f(G) \cap V)$;
then U_V is an invariant closed neighborhood of 1 in G for all $V \in \mathfrak{U}$. The
intersection of all U_V, $V \in \mathfrak{U}$ contains only the identity, because f is a continuous one-to-one representation; the filter basis of the U_V need not, however, be a neighborhood basis for 1. Let W' be a compact neighborhood of 1
containing $W^W = \cup \{ W^w : w \in W \}$. The filter basis of all compact neighborhoods $U_V \cap W'$, $V \in \mathfrak{U}$, intersects in the identity; hence there is a $V_o \in \mathfrak{U}$
such that $V \subset V_o$, $V \in \mathfrak{U}$ implies $U_V \cap W' \subset W$. Then $(U_V \cap W')^W \subset W^W \subset W'$
on one hand and $(U_V \cap W')^W \subset U_V^W \subset U_V$ on the other, so that $U_V \cap W' \subset$
$(U_V \cap W')^W \subset U_V \cap W'$. Thus the collection of all $U_V \cap W'$, $V \subset V_o$, $V \in \mathfrak{U}$ is
a basis of invariant neighborhoods of 1 in G since invariance under W
implies invariance under G, because W generates G.

(v) Let U be a compact invariant neighborhood of 1 in G and let U_V be
defined as in the proof of (iv); then the collection of all $U_V \cap U$, $V \in \mathfrak{U}$
clearly is a basis of invariant compact neighborhoods of 1.

The following two propositions state results of Freudenthal, Weil and
Iwasawa:

12.3. PROPOSITION. (*Freudenthal, Weil*). *A locally compact connected group
has property* (C) *if and only if it is the direct product of a vector subgroup and
a connected compact group.*

PROOF. [6; 23].

12.4. PROPOSITION. (*Iwasawa*). *A connected locally compact group satisfies*
(A) *if and only if its (topological) commutator group is compact.*

PROOF. [10].

12.5. EXAMPLE. The preceding proposition no longer holds for non-
connected groups: Let G be the unique nontrivial holomorphic extension
of the real additive group R by the group of two elements whose generator
maps r into $-r$ on R. Then every symmetric interval around 0 is invariant.
The commutator group, however, is R.

12.6. PROPOSITION. (*Iwasawa*). *Let G be a connected locally compact group with a compact invariant neighborhood of the identity. Then G contains the following normal subgroups:*

(i) *a unique maximal compact connected subgroup C,*

(ii) *a solvable connected subgroup S in the radical R of G such that the following statements are true:*

(iii) $G = SC$,

(iv) $S \cap C$ *is the commutator subgroup of R; it is central in G and $G/(S \cap C)$ is maximally almost periodic (hence has property* (B)),

(v) $S/(S \cap C)$ *is a vector space and $C/(C \cap S)$ is the maximal compact subgroup of $G/(S \cap C)$.*

PROOF. All this (and more) is explicitly or implicitly stated and proved in Iwasawa's paper [10]. Although we shall not need the full information, we try to illustrate Iwasawa's results in the following Hasse diagram:

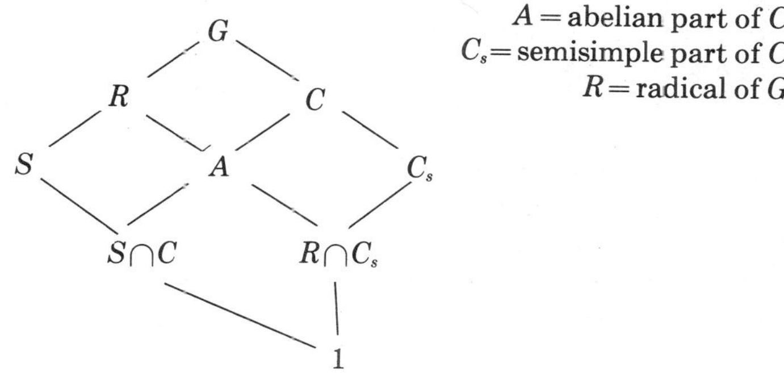

For the proof of some of the corollaries to our main theorem we need some simple lemmas:

12.7. LEMMA. *Let G be a vector group and $\mathfrak{A}$ a finite group of automorphisms of G. Then there exists a finitely generated discrete subgroup D of G which is invariant under $\mathfrak{A}$, and is such that G/D is compact.*

PROOF. If n is the dimension of G, then any finitely generated subgroup D of G has at most n independent generators and is, therefore, discrete. If it contains, in fact, n independent vectors, then G/D is compact.

Thus we have to show the existence of a group generated by a finite subset of G which contains n linear independent vectors and is permuted by $\mathfrak{A}$. If B is a basis for G, then $\{\alpha(b): b \in B, \alpha \in \mathfrak{A}\}$ is such a set.

12.8. **Lemma.** *Let G be a vector group and $\mathfrak{A}$ a finite group of automorphisms of G. Then, given any element g of G other than 1, there is a discrete subgroup D_g not containing g, invariant under $\mathfrak{A}$, and such that G/D_g is compact.*

Proof. Let D be as in Lemma 12.7. Let D_n be the subgroup of all n-th powers of elements in D. Given any element $g \in G$, then n can be made so large that g is not a point of D_n.

Theorem XII. *Let G be a locally compact group satisfying the following two conditions:*

(i) G/G_o is compact (G_o is the connected component of 1 in G).

(ii) G has an invariant compact neighborhood.

Then:

(I) G contains a compact subgroup C and a solvable connected subgroup S such that $S/(S \cap C)$ is a vector group and $G/(S \cap C)$ is a semidirect product of the normal vector subgroup $S/(S \cap C)$ and the compact group $C/(S \cap C)$; moreover $S \cap C$ is the commutator group of R, the radical of G_o, and is in the center of G_o.

(II) C contains a normal subgroup C_1 of finite index containing $S \cap C$. Let $G_1 = SC_1$. Then $G_1/(S \cap C)$ is the direct product of the vector group $S/(S \cap C)$ and the compact group $C_1/(S \cap C)$. (Clearly G_1 has finite index in G.) Moreover, $S \cap C$ is the intersection of all compact invariant neighborhoods of 1 in G, and $G/(S \cap C)$ is maximally almost periodic and has small invariant neighborhoods.

(III) There exist one-dimensional vector subgroups $N_1, \cdots, N_r$ in G, pairwise intersecting in the identity, such that:

(a) $G_1 = C_1 N_1 \cdots N_r$, $r = \dim S/(S \cap C)$,

(b) $C_1 N_{i_1} \cdots N_{i_t}$ is normal in G_1 for $1 \leqq i_1, \cdots, i_t \leqq r$,

(c) C_1 commutes element-wise with all N_i,

(d) $S \cap C$ is in the center of G_1.

If, conversely, a locally compact group satisfies (I) and (II), then it satisfies (i) and (ii).

The significance of the theorem may be illustrated by the following diagram which contains some additional information:

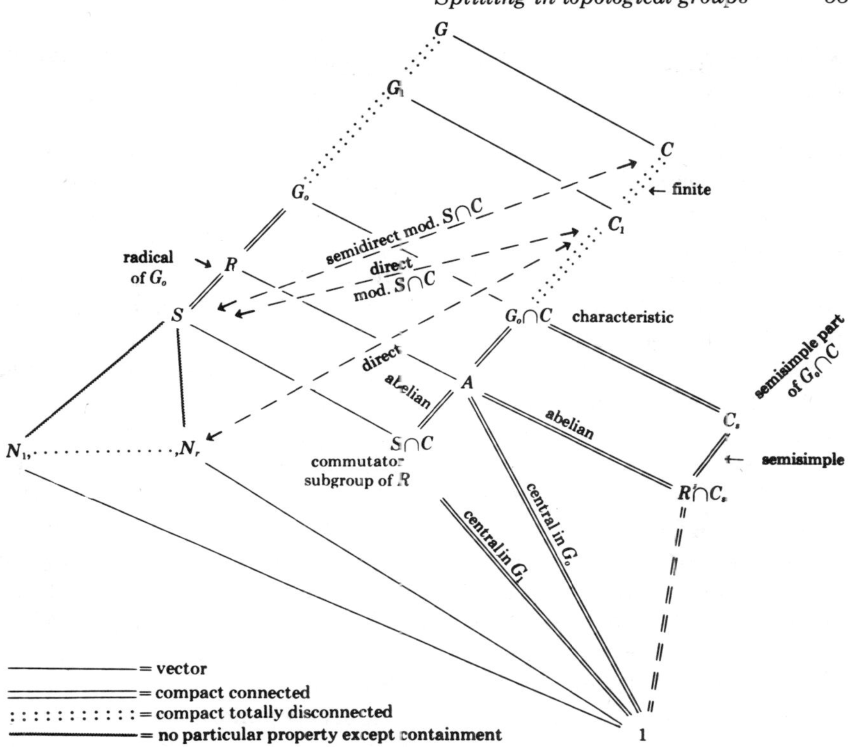

PROOF. (a) Let C_o be the compact subgroup of the component G_o and S_o the solvable group mentioned in Proposition 12.6. Then $S_o \cap C_o$, as the commutator group of the radical of G_o, is characteristic and hence normal in G. $G_o/(S_o \cap C_o)$ is a closed subgroup of $G/(S_o \cap C_o)$ and is a direct product of a vector group $S_o/(S_o \cap C_o)$ and the compact group $C_o/(S_o \cap C_o)$. Hence, by Corollary X.3., $G/(S_o \cap C_o)$ is a semidirect product of a vector group $S/(S_o \cap C_o)$ isomorphic to $S_o/(S_o \cap C_o)$ and a compact group $C/(S_o \cap C_o)$ containing $C_o/(S_o \cap C_o)$. Obviously $S \cap C = S_o \cap C_o$; and S is solvable since $S/(S \cap C)$ is abelian and $S \cap C$ is abelian. C is compact since $S \cap C$ and $C/(S \cap C)$ both are compact. This proves I.

Let G_1 be a subgroup of G containing G_o such that $G_1/(S \cap C)$ is the centralizer of $S/(S \cap C)$, and let C_1 be $G_1 \cap C$. Clearly, C_1 contains C_o since $C_o/(S \cap C)$ commutes element-wise with $S/(S \cap C)$. The quotient

$$(G/(S \cap C))/(G_1/(S \cap C))$$

which is isomorphic to G/G_1 and to C/C_1 acts effectively on the vector group $S/(S\cap C)$ according to 7.1.; it is, however, as a homomorphic image of G/G_o, totally disconnected. By 7.2. it acts as a closed subgroup of an orthogonal group on the vector group $S/(S\cap C)$ with respect to a suitable positive definite bilinear form. But any closed totally disconnected subgroup of an orthogonal group is discrete and therefore finite. This shows that G/G_1 and C/C_1 both are finite. The centralizer of a normal subgroup is clearly normal so that C_1 indeed is normal. Thus the product $S/(S\cap C)\cdot C_1/(S\cap C)$ is direct. By Lemma 12.8. for any element $s(S\cap C)$ in $S/(S\cap C)$, different from $S\cap C$, there is a discrete subgroup $D_s/(S\cap C)$ in $S/(S\cap C)$ invariant under the action of the finite group $G/(S\cap C)/(G_1/(S\cap C))$ and hence normal in $G/(S\cap C)$, and $D_s/(S\cap C)$ does not contain $s(S\cap C)$, but

$$(S/(S\cap C))/(D_s/(S\cap C))$$

is compact; hence $(G/(S\cap C))/(D_s/(S\cap C))$ is compact. Hence $G/(S\cap C)$ admits a continuous homomorphic image into a compact group such that any element of the form $sc(S\cap C), c\in C$ is not mapped on the identity. Since s was arbitrary (apart from $s\notin C$) and since every element of $G/(S\cap C)$ is of the form $sc(S\cap C)$, with some $s\in S$ and $c\in C$, we have proved that $G/(S\cap C)$ admits a one-to-one representation into a compact group. Therefore it is maximally almost periodic. Since G/G_o is compact and G_o, as a connected group, is generated by a compact neighborhood of the identity, G is generated by a compact neighborhood of the identity. Therefore $G/(S\cap C)$ has small invariant neighborhoods: 12.2.(iv). Thus $S\cap C$ contains the intersection of all invariant neighborhoods in G. But even in G_o the subgroup $S\cap C$ was the smallest subgroup such that $G_o/(S\cap C)$ had small invariant neighborhoods; hence, a fortiori, $S\cap C$ is the smallest subgroup of G such that $G/(S\cap C)$ has small invariant neighborhoods. This finishes the proof of II.

If G is a locally compact group satisfying I and II, then G/G_o is isomorphic to C/C_o and is, therefore, compact. As we have just shown, $G/(S\cap C)$ is maximally almost periodic and has small invariant neighborhoods; the counter image in G of an invariant compact neighborhood of the identity in $G/(S\cap C)$ is a compact invariant neighborhood in G. Thus the properties (i) and (ii) on one hand and (I) and (II) on the other are equivalent.

We have to prove (III): Suppose that (a) and (c) are satisfied; then C_1N_i is, for $1\leq i\leq r$, a direct product and all elements of C_1 commute with all elements on the groups N_i. Furthermore, N_i commutes with N_j element-wise modulo C_1, and thus (b) is satisfied. The center $S\cap C$ of S is certainly in the center of G_1. Hence, (b) and (d) of III are proved. It remains to verify (a) and (c). We prove this by induction on the dimension of the vector group $S/(S\cap C)$. If this dimension is zero, then (a) and (c) are trivial. Suppose that the statement is true for $\dim S/(S\cap C)=n$. Let E contain $S\cap C$ and be such that $E(S\cap C)$ is a one-dimensional vector subspace of $S/(S\cap C)$. Let F be such that $F/(S\cap C)$ is a supplementary $n-1$-dimensional subspace of $S/(S\cap C)$. By the induction hypothesis, $FC_1=C_1N_1'\cdots N_{n-1}'$, and $EC_1=C_1N$; by Proposition 9.4. we may assume that C_1N is a direct product, hence $G_1=EFC_1=EC_1N_1'\cdots N_{n-1}'=C_1NN_1'\cdots N_{n-1}'$; this finishes the proof by induction.

12.9. **EXAMPLE.** Let $G_o=G(V_1/Z_1,V_2,\beta_1)$ be the group of Example 11.9. The mapping α: $(x,(y,z))\to(-x,(-y,z))$ is an involutive automorphism (i.e. an automorphism of order 2) of G_o. Let S_n be the symmetric group on n elements and A_n the alternating group on n elements, a subgroup of S_n of index 2. We denote by ϕ the homomorphism of S_n into the automorphism group of G_o which maps all elements of A_n on the identity automorphism and all elements of $S_n\backslash A_n$ onto α. Let G be the holomorph of G_o with S_n over ϕ, i.e. the product $G_o\times S_n$ with the multiplication $(g,s)(h,t)=(g^{\phi(t)}h,st)$. Then:
- (i) $S=G_o\times\{1\}$,
- (ii) $C=G(V_1/Z_1,0,\beta_1)\times S_n$,
- (iii) $C'=G(V_1/Z_1,0,\beta_1)\times A_n$,
- (iv) $N_1=G(\pi(0),V_1\times\{0\},\beta_1)\times\{1\}$, with $V_2=V_1\times V_1,\pi:V_1\to V_1/Z_1$,
- (v) $N_2=G(\pi(0),\{0\}\times V_1,\beta_1)\times\{1\}$.

Compare also 11.11. which satisfies the conditions of the theorem.

12.10. **LEMMA.** *Let G be a locally compact group. Then the left uniformity is equal to the right uniformity if and only if G has small invariant neighborhoods.*

PROOF. If right and left uniformity coincide, then for each neighborhood U of 1 there is a neighborhood V of 1 such that $gV\subset Ug$ or $gVg^{-1}\subset U$ for every g in G. Then $V'=\bigcup\{gVg^{-1}:g\in G\}$ is an invariant neighborhood contained in U.

If $\mathfrak{U}$ is a neighborhood basis for 1 consisting of invariant neighborhoods U. Then the collection of sets $\bigcup\{Ug\times\{g\}:g\in G\}$ is a basis for the right and left uniformity of G.

COROLLARY XII.1. *Let G be a locally compact group such that G/G_o is compact. Then the following statements are equivalent:*
- (i) *G is maximally almost periodic,*
- (ii) *G has sufficiently many finite dimensional unitary representations,*

(iii) *G has sufficiently many strongly continuous unitary representations into rings of finite type* (I_n *and* II_1),

(iv) *G has sufficiently many unitary representations continuous in the uniform operator topology,*

(v) *G has small invariant neighborhoods,*

(vi) *The right and left uniformities of G are the same,*

(vii) *G is the semidirect product of a normal vector subgroup N and a compact subgroup C, and C contains an open subgroup C' normal in G of finite index in C such that NC' is a direct product. If G is also a semidirect product of N and D, then C and D are conjugate.*

PROOF. We observe that G is generated by a compact neighborhood of the identity, because G/G_o is compact.

(i) and (ii) are equivalent: 12.2.(ii),

(v) and (vi) are equivalent: 12.10,

(i) implies (v): 12.2.(iv),

(v) implies (vii): Theorem XII, and, as far as conjugacy is concerned, Corollary X.3,

(vii) implies (i): Theorem XII.

Thus, (i), (ii), (v), (vi), and (vii) are equivalent.

(ii) implies (iii) and (iv): Trivial.

(iii) or (iv) implies (vii): This has been proved by Kadison and Singer in the case that $G=G_o$. That is, the component G_o is a direct product of a vector group and a compact group. Then, Corollary X.3. immediately proves (vii). This closes the circle.

The history of the investigation of the class of groups described by Corollary XII.1 in the connected case suggests an attribution of the conditions to several authors: (i) to Freudenthal [6], (ii) to Weil [23], (iii) and (iv) to Kadison and Singer [11], (v) to Pontrjagin, van Kampen [12], Mostow [17], Iwasawa [10]; condition (vi) was used by Braconnier [3] to characterize locally compact groups having this property as direct products of vector groups and compact groups; the error in this statement was pointed out by Freudenthal (see Mathematical Reviews, vol. 18(1957), p.745). The semidirect product structure in. (vii) appears in a paper by Murakami [27]. Kuranishi [26] gave an example of a totally disconnected maximally almost periodic locally compact group without any compact open normal subgroups.

The following theorem was announced by van Kampen [12, p. 463], is mentioned in [22, p. 129] and should have taken the place of Theorem XIII in the present outline:

"If G is a locally compact group with small invariant neighborhoods, then G is the direct product of a vector subgroup N and a group H which contains a compact open normal subgroup."

The hypothesis added l.c. that G be separable metric was necessary since the characterization of not necessarily separable connected locally compact maximally almost periodic groups was provided by Weil not before 1938; the statement of van Kampen's theorem in Weil's book was originally given with the weaker hypothesis that G should contain at least one invariant neighborhood and with the additional hypothesis that G be generated by a compact neighborhood of the identity. The statement that one compact invariant neighborhood is sufficient was corrected in the second edition due to the work of Iwasawa [10] and Mostow [17].

That van Kampen's theorem is incorrect even with the additional assumption is exhibited by Example 11.10. which is the Lie group M defined on the cartesian product of the real line R and the product $Z \times Z$ of two infinite cyclic groups by the multiplication $(x,y,z)(x',y',z') = (x+x' + (yz' - y'z), y+y', z+z')$. This is a Lie group generated by a compact neighborhood of the identity, has small invariant neighborhoods, a one-dimensional open vector subgroup which does not split, and M is maximally almost periodic.

If one adds to van Kampen's theorem the hypothesis that G/G_o is compact, then it becomes almost true, as Corollary XII.1. shows; thus we have, unfortunately, to dispense with Theorem XIII and to replace it by Corollary XII.1. We hope, however, that it is a contribution to the elucidation of this matter called for by Weil in the section of his book quoted above.

12.11. DEFINITION (Iwasawa). A locally compact group G is called a *C-group* (or is said to satisfy *condition C*) if the quotient G/R of G modulo its radical, i.e. its maximal solvable connected normal subgroup, is compact [9].

REMARK. This definition is slightly different from that given by Iwasawa in the non-connected case. However, it seems to be the more natural. In particular, it is equivalent to the definition that Iwasawa gives together with the assumption that G/G_o is compact.

12.12. DEFINITION (Takenouchi). Let G be a locally compact group, let $C_o(G)$ denote the space of all continuous complex valued functions on G with compact carrier, with the compact open topology, and $f*f \to C_o(G)$ the convolution defined for $f \in C_o(G)$ by $(f*f)(g) = \int_G f(gh)f(h)dh$ with the Haar integral on G. Then G is said to satisfy condition R if the closure of the set $\{f*\bar{f}:f \in C_o(G)\}$ in the space $C_o(G)$ of all continuous functions with the compact open topology contains the constant functions (i.e. the function 1 can be uniformly approximated on every compact set by functions $f*f$ [22].)

12.13. PROPOSITION (Takenouchi). *Let G be a locally compact group such that G/G_o is compact. Then G satisfies condition C if and only if it satisfies condition R.*

Proof. [22].

COROLLARY XII. 2. *Let G be a locally compact group such that G/G_o is compact. Then G is a C-group if and only if the following conditions are satisfied:*

(i) *G contains a maximal compact subgroup C.*

(ii) *The commutator series $R = R^{(0)} \supset R^{(1)} \supset \cdots \supset R^{(n)} = 1$ has a refinement to a normal series*

$$R \supset S^{(1)} \supset R^{(1)} \supset S^{(2)} \supset \cdots \supset S^{(n-1)} \supset R^{(n-1)} \supset S^{(n)} \supset 1$$

such that for $i = 0, \cdots, n-1$, $R^{(i)}C/R^{(i+1)}$ is the semidirect product of the normal vector subgroup $S^{(i)}/R^{(i+1)}$ and the compact group $CR^{(i+1)}/R^{(i+1)}$.

PROOF. Suppose that G is a C-group. We prove the corollary by induction on n. By Corollary X.3. the corollary is true for $n = 1$.

Suppose it has been proved for length $n-1$ and that R has length n. Denote the abelian subgroup $R^{(n-1)}$ with M; since M is characteristic in R, it is normal in G. Clearly G/M is a C-group and its radical R/M has length $n-1$. The induction hypothesis applies. Hence, there is a subgroup C_1 of G such that C_1/M is a maximal compact subgroup of G/M. $R^{(i)}C_1/R^{(i+1)}$ is the semidirect product of a vector group $S^{(i)}/R^{(i+1)}$ and the compact group $C_1R^{(i)}/R^{(i+1)}$ for $i = 0, \cdots, n-2$. The group C_1 contains the connected abelian normal subgroup M and C_1/M is compact. M is the direct product of a vector space $S^{(n)}$ and a compact group. By Corollary X.3. we may assume that C_1 is the semidirect product of $S^{(n)}$ and a compact group C. But because $R^{(i)}C_1 = R^{(i)}C$ and $C_1R^{(i)} = CR^{(i)}$ for $i = 0, \cdots, n-1$, the corollary is proved in one direction.

If, however, the conditions (i) and (ii) are satisfied, then it is easy to see that $G/R = RC/R$ is compact because C is compact and hence G is a C-group.

COROLLARY XII.3. *Let G be a locally compact maximally almost periodic group and suppose that G is generated by a compact neighborhood of the identity or else contains at least one invariant compact neighborhood. Then G contains* *an open normal subgroup G' which is a direct product of a vector group and a compact group.*

PROOF. In both instances, G has small invariant neighborhoods. Let G_1 be any open subgroup such that G_1/G_o is compact. Then, by Corollary XII.1, G_1 contains an open subgroup G_2 which contains a central vector subgroup N such that G_2/N is compact. Choose now a compact neighborhood of 1 in G_2, invariant in G. It generates a normal subgroup G' of G which contains N in its center such that G'/N is compact. Then, e.g. by Corollary X.3, G' is a semidirect, and, because of the centrality of N, a direct product of N and some compact group.

COROLLARY XII.4. *Let G be a locally compact or first countable topological group with connected component G_o and a closed central subgroup N, satisfying the following conditions:*

(i) *N is isomorphic to the additive group of a reflexive Banach space,*

(ii) *G/N is maximally almost periodic locally compact,*

(iii) *G/G_o is compact.*

Then there exist

(i') *a closed subgroup N_o of N isomorphic to a Hilbert group,*

(ii') *a nilpotent Lie group L with dimension at most $n(n+1)/2$ and homeomorphic to a euclidean space, where n is the dimension of G/N modulo a maximal compact subgroup,*

(iii') *a pair of compact subgroups C',C with $C'\subset C$, such that $G=N_oLC$, and the product N_oLC' is direct and of finite index in G. The subgroup N_oL is normal in G.*

PROOF. This corollary is an immediate consequence of Corollary XI.1. and Corollary XII.1.

The following corollary was announced in Section VIII and generalizes Iwasawa's lemma (Corollary VIII.1) and Corollary IX.2.

COROLLARY XII.5. *Let G be a locally compact or first countable topological group with connected component G_o and a closed normal subgroup N satisfying the following conditions:*

(i) *N is isomorphic to the additive group of a reflexive Banach space,*

(ii) *G/N is maximally almost periodic locally compact,*

(iii) *G/G_o is compact.*

Then there exists a normal closed subgroup S containing N which is vector solvable of length 2 and a maximal compact subgroup C such that G is the semidirect product of S and C.

PROOF. Observing Corollary XII.1. (vii), the proof is the same as the proof for Corollary IX.2.

In Section IX we use the result proved below which is independent of the remainder of this section; it belongs however in the general context of the present chapter and is of some independent interest. It is given in the following as an appendix to Section XII.

12.14. PROPOSITION. *Let G be a locally compact topological group which is the semidirect product of a compact normal subgroup C and a closed maximally almost periodic subgroup A. Let ϕ be the continuous representation of A into the automorphism group of C which assigns to $a \in A$ the automorphism $a \to a^{-1}ca$ of C. Suppose that the closure of $\phi(A)$ in the automorphism group of C (endowed with the compact open topology) is compact. Then G is maximally almost periodic.*

PROOF. First we note that G is isomorphic to the holomorph $C \otimes_\phi A$ defined on the product space $C \times A$ with the following multiplication:

$$(c,a)(c',a') = (cc'^{\phi(a)^{-1}}.aa').$$

Let A^* be the almost periodic compactification of A, i.e. A^* is a compact group having the following properties:

(i) there is a one-to-one representation f of A onto a dense subgroup of A^*,

(ii) any representation ψ of A into a compact group admits an extension to A^*, i.e. there is a representation ψ^* of A^* into the same group such that $\psi = \psi^* \circ f$. (See [23, p.125, 135].)

Let now ϕ^* be the extension of the representation ϕ of A into the compact group $\phi(A)$ to a representation of A^* into $\phi(A)$ with $\phi = \phi^* \circ f$. We form the holomorph $C \otimes_\phi A$ which is defined on the product space $C \times A$ by the following multiplication: $(c,a)(c',a') = (cc'^{\phi^*(a)^{-1}},aa')$. The mapping $(c,a) \to (c,f(a))$ is obviously a one-to-one representation of $C \otimes_\phi A$ into the compact group $C \otimes_\phi A$. Hence $A \times C$ and therefore G is indeed maximally almost periodic (see 12.2. (iii)).

XIII. SIXTH APPLICATION.

13.1. A locally compact group G is said to admit *two ends* if it can be compactified on adding two non-isolated points. (For the connected case see [7].)

13.2. PROPOSITION. (*Freudenthal*). *Let G be a locally compact connected group satisfying the first axiom of countability and admitting two ends. Then G is a direct product of a one-dimensional vector group and a compact connected group.*

PROOF. [7].
Although Freudenthal's method can be generalized to cover the non-separable case and a proof for the general case has been given by Iwasawa [10], we derive this result from the theorem quoted above.

13.3. PROPOSITION. *Let G be a locally compact connected group, admitting two ends. Then G is the direct product of a one-dimensional vector group and a compact group.*

PROOF. It is well known that G contains a compact normal subgroup H such that G/H satisfies the first axiom of countability [15, p.58]. If G admits two ends, then G/H obviously also admits two ends since H is compact. By Freudenthal's theorem (13.2), G/H is the direct product of a one-dimensional vector group N'/H and a compact group C'/H. Clearly C' is a compact normal subgroup of G, and G/C' is isomorphic to

$$(G/H)/(C'/H)$$

i.e. to a one-dimensional vector group. By Iwasawa's theorem [9, p. 549] (or by 14.3 of the following section) there is a one parameter group N in G such that G is the semidirect product of the normal subgroup C' and N. Then, by 12.13, N is also normal and the product is direct.

THEOREM XIV. *Let G be a locally compact group admitting two ends and let G_o be the connected component of G. Then the following statements are equivalent:*

(i) There is a two point compactification such that in it the closure of all cosets gG_o modulo G_o contains both points at infinity,

(ii) G/G_o is compact,

(iii) G contains a normal one-dimensional vector group N and a compact group C such that G is the semidirect product of N and C. Furthermore, C contains an open subgroup C' of index at most two in C such that NC' is a direct product and is of index at most two in G. If $G=NC=ND$ and these products are semidirect, then C and D are conjugate.

PROOF. It is trivial that (iii) implies (ii).
Suppose now, that (i) is satisfied. By 13.3, G_o is a direct product of a one-

dimensional vector group N and a compact group C_o. Since C_o is characteristic in G_o, it is normal in G, and since it is compact, G/C_o admits a two point compactification such that every component touches both points at infinity. We shall now show that $(G/C_o)/(G_o/C_o)$ is compact, which will imply that the group G/G_o isomorphic to it is compact, which is (ii). It is therefore sufficient to prove that (i) implies (ii) in the case that the component G_o is an arc joining $-\infty$ and $+\infty$. If U and V are open disjoint neighborhoods of $-\infty$ and $+\infty$ respectively, then $G\backslash(U\cup V)$ is compact and no arc joining $-\infty$ and $+\infty$ is entirely contained in the union of the two disjoint open sets U and V. Therefore G/G_o is a continuous image of the compact space $G\backslash(U\cup V)$ since each component gG_o meets $G\backslash(U\cup V)$; therefore gG_o is compact.

Suppose now that (ii) is satisfied. By 13.3, G_o is a direct product of a one-dimensional vector group N and a characteristic compact group C_o; thus G_o is contained in the centralizer N' of N. By Corollary X.3., we may suppose that G is the semidirect product of N and a compact subgroup C of G. If we set $C' = N' \cap C$, then C/C' acts as an effective compact group of automorphisms on N and is therefore at most of order two. It is obvious that NC' is a direct product since C' is in the centralizer of N. This finishes the proof.

13.4. **EXAMPLE.** Let R be the additive group of reals and Z the additive group of all integers. Then $R \times Z$ admits a connected two point compactification $R \times Z \cup \{-\infty, +\infty\}$ of the following kind:

Let $-\infty$, $+\infty$ be the two ends of $R \times \{0\}$, let all arcs $R \times \{2n\}$, $n \in Z$, $n \neq 0$, be one point compactified by $-\infty$ and all arcs $R \times \{2n+1\}$, $n \in Z$ by $+\infty$. Let each neighborhood of $-\infty$ (respectively $+\infty$) contain all but a finite number of the above arcs touching $-\infty$ (respectively $+\infty$). This example illustrates the rôle of condition (i) above.

13.5. **EXAMPLE.** Let R be the additive group of reals. Let S_n be the full group of permutations of $n > 2$ elements, and A_n the subgroup of all even permutations. Let $\phi : S_n \to \mathfrak{A}(R)$ be the homomorphism of S_n into the automorphism group $\mathfrak{A}(R)$ of R such that all even permutations leave R fixed and that all odd permutations reflect R at 0. Clearly A_n is the kernel of ϕ. Let $G = R \otimes S_n$ be the holomorph of R with S_n under the action of ϕ, i.e. the multiplication is defined by

$$(r,p)\,(s,q) = (r^{\phi(q)}s, pq).$$

Then $R \times A_n$ is a direct product but $\bar{R} \times A_n$ does not split in G. This example shows that the structural analysis in (iii) above cannot be improved.

COROLLARY XIV.1. *Let G be a locally compact group whose connected component is G_o. Suppose that the following two conditions are satisfied:*
(i) G_o is not compact,
(ii) there is an element $g \in G$ such that, with the cyclic subgroup $G(g)$ generated by g, the quotient space $G/G(g)$ is compact Hausdorff.
Then G contains a normal one-dimensional vector group N and a compact group C such that G is the semidirect product of N and C. Moreover, C contains an open subgroup C' of index at most two such that NC' is a direct product and is of index at most two in G. If $G = ND$, D compact, then C and D are conjugate.

PROOF. (a) There exists a neighborhood K of 1 with compact closure such that $G = KG(g)$ (i.e. G is syndetic); for if V is a compact neighborhood of 1 in G, then, by (i), a finite number of translates $g_1 VG(g), \cdots, g_m VG(g)$ cover G, the union K of the sets $g_1 V, \cdots, g_m V$ satisfies the requirements.

(b) G admits a two point compactification: Clearly, $G(g)$ is infinite because G is not compact. Let $-\infty$ and $+\infty$ be two different points not contained in G. We define a basis of the neighborhoods of $-\infty$, respectively $+\infty$, to be a collection of sets of the form

$$U_n^- = \{-\infty\} \cup \bigcup \{Kg^{-k} : k > n\} \qquad n = 1, 2, \cdots,$$

$$U_n^+ = \{+\infty\} \cup \bigcup \{Kg^k : k > n\} \qquad n = 1, 2, \cdots.$$

Let the topology of $G \cup \{-\infty, +\infty\}$ be generated by the topology of G and these sets. It is immediately clear that this topology induces the given topology on G, and that G is dense. This topology is Hausdorff: It is easy to see that K meets only a finite number of sets Kg^t, $t \in F$; let $n = \max\{|t| : t \in F\}$; then $U_n^- \cap U_n^+ = \emptyset$. Since the complement of $U_r^- \cup U_s^+$ is compact for all $r, s = 1, 2, \cdots$, the space $G \cup \{-\infty, +\infty\}$ is indeed compact.

(c) G/G_o is compact: Let G_1 be an open subgroup such that G_1/G_o is compact. It is sufficient to show that some power g^n is in G_1; then G/G_1 is a continuous image of $G/G(g^n)$ which is compact. The image of $G(g)$ in G/G_o cannot be infinite discrete, for then the product $G_oG(g)$ would be semidirect and G_o would be mapped homeomorphically into $G/G(g)$ which is impossible because G_o is not compact. Hence in the neighborhood G_1/G_o of the identity in G/G_o we find an element G_og^n of $G_oG(g)$ different from the identity. This proves the claim.

The corollary now follows from Theorem XIV.

REMARK. If, in addition to the hypotheses of the corollary, G does not contain compact normal subgroups, then G is either isomorphic to the reals or to the nontrivial holomorphic extension of the reals by the group of two elements. If G does not contain compact subgroups, then it is isomorphic to the reals. The latter result has recently been proved by Chu [5]. A portion of the corollary has been obtained by one of the authors [8].

If, in the corollary the hypothesis (i) is replaced by (i') G_o is compact, then the following is known: Chu proves, l.c., that G, provided it has no compact subgroups, is infinite cyclic discrete. In [8] it is proved that G, provided that the multiplication of G can be continuously extended to $+ \infty$, is the semidirect product of a maximal compact normal subgroup and an infinite cyclic group. It may be conjectured that a locally compact group G satisfying (i') and (ii) has an open subgroup G' of index at most two, which is a semidirect product of a compact normal group and an infinite cyclic group.

XIV. UNIMODULARITY.

14.1. LEMMA. *Let G be a Lie group and H a normal closed subgroup such that G/H is isomorphic to the additive group of the reals. Then G contains a one-parameter subgroup E isomorphic to G/H such that G is the semidirect product of H and E.*

PROOF. The Lie algebra $\mathfrak{h}$ of H is an $(n-1)$-dimensional ideal in the n-dimensional Lie algebra $\mathfrak{g}$ of G. Let e be any element of $\mathfrak{g}$ not contained in $\mathfrak{h}$; the one-parameter group E generated by e is the required one.

14.2. LEMMA. *Let G be a locally compact connected group and H a closed normal subgroup such that G/H is isomorphic to the additive group of the reals. Then G contains a one-parameter subgroup E isomorphic to G/H such that G is the semidirect product of H and E.*

PROOF. Let C be a compact normal subgroup of G such that G/C is a Lie group. Necessarily C is contained in H since G/H does not contain compact subgroups other than the identity. Therefore G/C satisfies the conditions of Lemma 14.1. and contains a subgroup E'/C which is isomorphic to the additive reals such that G/H is the semidirect product of E'/C and H/C. Now E' is a two-ended group since E'/C is obviously two-ended and C is compact. Hence E' is the product of a one-parameter group E and C (Theorem XIV). Clearly E satisfies the requirements.

14.3. LEMMA. *Let G be a locally compact group and H a closed normal subgroup such that G/H is isomorphic to the additive group of reals. Then G contains a one-parameter group E isomorphic to G/H such that G is the semidirect product of H and E.*

PROOF. Let G_o be the connected component of 1. Evidently G_o and $G_o \cap H$ satisfy the hypotheses of Lemma 14.2. Hence G_o is the semidirect product of $(G_o \cap H)$ and of a one-parameter group E isomorphic to G/H. The group HE/H is a nontrivial connected subgroup of G/H, which is isomorphic to the reals, therefore $G/H = HE/H$ and $G = HE$. Since E is evidently mapped one-to-one, $H \cap E = \{1\}$. Hence the product HE is semidirect.

14.4. PROPOSITION. *Let G be a locally compact group and Δ its unimodular factor (see 5.4). Let $\Delta(G)$ be the full group of positive reals. Then G is a semidirect product of its maximal unimodular subgroup U (which is normal in G) and a one-dimensional vector subgroup N. The group U is not compact.*

PROOF. U is the kernel of Δ [23, p.44] and G/U is isomorphic to the additive reals. Then Lemma 14.3. applies. If U were compact, then the product UE would be direct by Proposition 9.4. Then G would be unimodular which is not true.

14.5. PROPOSITION. *Let G be a locally compact group and G_o its connected component. Suppose that G/G_o is compact. Then either G is unimodular or a semidirect product of its maximal unimodular subgroup U and a one-dimensional vector subgroup N, in which case U cannot be compact.*

PROOF. Let Δ be the unimodular factor of G. If $\Delta(G_o)$ is not trivial, then it is a connected subgroup of the group of all positive reals under multiplication and is therefore all of it, in which case Proposition 14.4 applies. If $\Delta(G_o)$ is trivial, then G_o is connected in U and hence G/U is compact. But the only compact subgroup contained in the positive reals is $\{1\}$. Hence $G = U$ and U is unimodular.

14.6. PROPOSITION. *Let G be a locally compact group and U its maximal unimodular subgroup. Then either U is open in G or G is the semidirect product of U and a one-dimensional vector subgroup N.*

PROOF. Suppose that $\Delta(G_o)=\{1\}$. If G' is an open subgroup such that G'/G_o is compact, then $\Delta(G')=\{1\}$. Hence $G'\subset U$ which proves that U is open.

THEOREM XV. *Let G be a locally compact group and U its maximal unimodular subgroup (which always exists and is normal). Then one and only one of the following cases is true:*

(i) *$G=U$, i.e. G is unimodular,*

(ii) *G is the semidirect and not direct product of U and a one-dimensional vector subgroup N, and U is not compact,*

(iii) *U is open and G/U is a discrete abelian group algebraically isomorphic to a subgroup of the reals (hence it is torsion free and contains no nontrivial compact subgroups).*

PROOF. The proof follows immediately from the preceding propositions.

14.7. **EXAMPLE.** Let R be the additive reals and R_o be the positive multiplicative reals with the discrete topology. Introduce the following multiplication on $G=R\times R_o$: $(r,s)(r',s')=(s'r+r',ss')$ (i.e. G is a holomorph of R by R_o). Then $U=R\times\{0\}$ is the maximal unimodular subgroup. The measure defined by the length of an interval $I\times\{s\}$ is a left invariant measure since $(r,s)I\times\{s'\}=(s'r+I)\times\{ss'\}$; it is, however, not right invariant:

$$I\times\{s\}(r',s')+(s'I+r')\times\{ss'\}.$$

Clearly any subgroup of R_o renders the same service. If R_o is given the natural topology, we get an example illustrating case (ii) of the theorem.

14.8. **EXAMPLE.** Let R_o be the integers; form G with this group according to 14.7. Let $H=R\times G$ be the direct product of R and G. The subgroup D generated by $(1,0,0)$ and by $(0,1,0)$ is normal in H. The factor group H/D is an extension of a torus by a cyclic group. The torus is the maximal unimodular subgroup. One can, therefore, not conclude that in case (iii) U must not be compact.

BIBLIOGRAPHY

1. R. Baer, *Crossed isomorphisms*, Amer. J. Math. **64**(1944), 341—404.

2. ———, *Splitting endomorphisms*, Trans. Amer. Math. Soc. **61**(1947), 508—516.

3. J. Braconnier, *Remarques sur les groupes localement compacts dont les structures uniformes droite et gauche sont egales*, Ann. Univ. Lyon. Sect. A(3) **18**(1955), 15—19.

4. P. Cartier, *Structure topologique des groupes de Lie generaux*, Seminaire "Sophus Lie", E.N.S., (1954—55), Expose no. 22.

5. H. Chu, *A characterisation of integer groups and real groups*, Duke Math. J. **28**(1961), 57—77.

6. H. Freudenthal, *Topologische Gruppen mit genügend vielen fastperiodischen Funktionen*, Ann. of Math. **13**(1951), 57—77.

7. ———, *La structure des groupes á deux bouts et des groupes triplement transitifs*, Nederl. Akad. Wetensch. Indag. Math. **13**(1951), 288—294.

8. K. H. Hofmann, *Locally compact semigroups in which a subgroup with compact complement is dense*, Trans. Amer. Math. Soc. **106**(1963), 19—51.

9. K. Iwasawa, *On some types of topological groups*. Ann. of Math. **50**(1949), 507—558.

10. ———, *Topological groups with invariant neighborhoods of the identity*, Ann. of Math. **54**(1951), 345—348.

11. R. V. Kadison and I. M. Singer, *Some remarks on representations of connected groups*, Proc. Nat. Acad. Sci. U.S.A. **38**(1952), 419—423.

12. E. R. van Kampen, *Locally bicompact abelian groups and their character groups*, Ann. of Math. **36**(1935), 448—463.

13. J. L. Kelley, *General Topology*, Van Nostrand, Princeton (1955).

14. E. Michael, *Convex structures and continuous selections*, Canad J. Math. **11**(1959), 556—575.

15. D. Montgomery and L. Zippin, *Topological Transformation Groups*, Intersec. Tracts I(1955).

16. P. S. Mostert, *Sections in principal fibre spaces*, Duke Math. J. **23**(1956), 57—72.

17. G. D. Mostow, *On an assertion of Weil*, Ann. of Math. **54**(1951), 339—344.

18.   M. A. Naimark, *Normed Rings,* Groningen (1959). Transl. from the Russian, by P. Noordhoff.

19.   L. S. Pontrjagin, *Topologische Gruppen,* Teil 2. Übersetzung aus dem Russischen bei Teubner, Leipzig (1957).

20.   N. Steenrod, *The Topology of Fibre Bundles,* Princeton Univ. Press Princeton, N. J. (1951).

21.   T. E. Stewart, *Lifting group actions in fiber bundles,* Ann of Math. **74**(1961), 192—198.

22.   O. Takenouchi, *Sur une classe de fonctions continues de type positif sur un groupe localement compact,* Math. J. Okayama Univ. **3**(1953/54), 143—173.

23.   A. Weil, *L'integration dans les Groupes Topologiques et ses Applications,* Actualités Sci. Ind. **1145**(1951) = 869 (1940).

24.   H. Zassenhaus, *The Theory of Groups,* Transl. from the German by Chelsea, N. Y. (1949).

25.   L. Zippin, *Two ended topological groups,* Proc. Amer. Math. Soc. **1**(1950), 309—315.

26.   M. Kuranishi,[1] *On non-connected maximally almost periodic groups,* Tôhoku Math. J. **2**(1950), 40—46.

27.   S. Murakami,[1] *Remarks on the structure of maximally almost periodic groups,* Osaka Math. J. **2**(1950), 119.129.

[1]These two papers first escaped the author's attention. They belong to the context of Corollary XII.1.

Appendix

The theorem that we prove here should, more appropriately, succeed Section IX. However, the authors were unable to find the proof until after the main body of the manuscript had gone to press. Hence, we present it as an appendix.

THEOREM XVI. *If a locally compact group G contains a vector subgroup V such that the quotient space G/V is compact, then there is a closed normal vector subgroup N isomorphic to V and a compact subgroup C such that $G=NC$, $N \cap C = \{1\}$.*

This is evidently a generalisation of Iwasawa's theorem (Corollary VII.1). A group which is compact modulo its component has always the structure indicated in the theorem provided it is maximally almost periodic (Corollary XII.1), an observation which was used to prove this theorem in the case that V is one-dimensional (see Section XIII).

The theorem is first proved for solvable groups. In order to extend it to the general case we show that the factor group of G modulo the radical of the component of G is compact. For this important step we use a result of Borel which was also proved by Weil[1].

Recall that the radical R of a locally compact group G is the maximal connected solvable normal subgroup. We will call G *semisimple* if R is trivial. If G is compact, then semisimplicity is equivalent to the lack of nontrivial connected compact abelian groups central in the component of the identity.

As we know, a vector group in a topological group is a subgroup isomorphic to the additive group of a finite dimensional vector space over the field of the reals. Because of their completeness vector groups are closed in whatever groups they are imbedded.

The authors have learned through verbal and written communications of alternative proofs to Theorem XVI, in the case of semisimple connected Lie groups from J. Tits and in the case of connected Lie groups from

1 A. Borel, *Density properties for certain subgroups of semisimple groups without compact component*, Ann. of Math. **72**(1960), 179–188.

A. Weil, *On discrete subgroups of Lie groups* (II), Ann. of Math. **75**(1962), 578–602.

H. Freudenthal. It is well conceivable that there are shorter and more sophisticated proofs of this result. It seems, however, that some propositions proved in the course of our demonstration are of independent interest in the context of this memoir; thus we feel justified to present our original proof.

Solvable Groups

Lemma 1. *Let N be a closed normal vector subgroup of a topological group G with abelian compact factor group G/N. Then G' is a subvector group of N.*

Proof. Since G/N is abelian, G' is contained in N. Let A be a compact group isomorphic to G/N so that $NA = G$. Now G' is the closure of the subgroup generated by all elements $c(n_1,n_2,a_1,a_2) = (n_1a_1)^{-1}(n_2a_2)^{-1}n_1a_1n_2a_2$, $n_i \in N$, $a_i \in A$, $i = 1,2$. For each pair (a_1,a_2) the image of the connected space $N \times N$ under the continuous mapping $(n_1,n_2) \to c(n_1,n_2,a_1,a_2)$ is connected and contains 1. Thus the union of all these images is connected and so is the group generated by this union.

Lemma 2. *Let H be a closed normal subgroup of the topological group G such that G/H is compact. Let V be a vector group normal in H such that H/V is compact abelian. Then there is a vector subgroup N of H normal in G such that G/N is compact.*

Proof. By Corollary X.3, V splits in H as a semidirect normal factor. Then, by Lemma 1, H' is a vector subgroup of V. Since H' is characteristic in H, hence normal in G, the centralizer Z of H' in G is normal in G. Because V is contained in Z it is in $Z \cap H$. The abelian factor group $(Z \cap H)/H'$ is the direct product of the vector group V/H' and some compact abelian group C/H', where C is normal in $Z \cap H$. Now C/H' is characteristic in $(Z \cap H)/H'$ and hence C is even normal in G. Again by Corollary X.3, H' splits as a normal semidirect factor in C, i.e. $C = H'K$ with some compact abelian subgroup K of C. But H' is in the center of Z, thus C is abelian and K is characteristic in C, consequently normal in G. Now $Z \cap H = VK$ is a direct product and by a final application of Corollary X.3. we find a vector group N in $Z \cap H$ which is normal in G such that the factor group G/N is compact.

Proposition 3. *If a group G is the product of a normal vector group N and a vector group V, and if G/V is compact, then G is abelian.*

PROOF. The factor group $N/(N \cap V)$ is homeomorphic to the quotient space $NV/V = G/V$; the latter is compact, thus $N \cap V$ must contain a free abelian group F of the same rank as N. The inner automorphisms with elements of V induce linear transformations on the vector group N. But since they leave F elementwise fixed, they act trivially on N. Thus V is in the centralizer of N which proves the proposition.

LEMMA 4. *Let G be a locally compact group such that G_o is solvable. If G contains a vector subgroup such that the quotient space G/V is compact, then there is a normal vector subgroup N such that G/N is compact.*

PROOF. A maximal compact connected subgroup M of G_o is solvable, hence abelian. Now G_o is the product of the two abelian subgroups V and M. Therefore $G'' = 1$ by a theorem of Itô[2]. Now let C be a maximal compact connected subgroup of G'_o. It is characteristic in the abelian group G' and is, therefore, normal in G_o. In particular, C is normal in VG'_o. Now VG'_o/C is the product of the normal vector group G'_o/C and the vector group VC/C. Hence it is abelian by Proposition 3. So it is the direct product of the vector group VC/C which has the largest possible dimension and some compact connected group K/C where K is normal in VG'_o and is compact and connected. Now $VG'_o = VK$; the automorphism group of K is totally disconnected ([9], p. 509), hence the action of the connected group V under inner automorphisms on K must be trivial. Therefore VK is a direct product. Because G_o/G'_o is abelian, VG'_o is normal in G_o. Then, again by Corollary X.3. there is a vector group W in VG'_o normal in G_o such that the factor group G_o/W is compact. As a compact connected solvable group, this factor group is abelian. Thus the hypotheses of Lemma 2 are satisfied which proves the remainder.

PROPOSITION 5. *Let N be a normal vector group in a locally compact group such that G/N is a vector group. Suppose further that A is an abelian closed normal subgroup of G containing N such that G/A is compact. Then N splits in G as a semidirect factor. If, in particular, N is central, then G is a vector group.*

PROOF. Let A_o be the component of A; then A_o is a vector group normal in G. If A_o splits in G so does N, for according to 7.1 the compact group G/A acts on N as well as on A_o as a compact group of automorphisms. Hence by

2 N. Itô, *Über das Produkt von zwei abelschen Gruppen*, Math. Z. **62**(1955), 400–401.

7.2 and 7.3 A_o decomposes into a direct product of N and a normal subgroup N_1. Hence if A_o splits in G so does N. Henceforth we may assume $N = A_o$. Then A is the direct product NF of N and a discrete free abelian group F. We show that F may be chosen normal in G: The compact group G/A acts on NF under inner automorphisms (cf. Lemma 7.1.). Let $\phi: NF \to R^n$ be an embedding into n-dimensional real vector space where $n = \operatorname{rank} N + \operatorname{rank} F$. Obviously G/A acts as a group of automorphisms on $\phi(NF)$ and each one of these automorphisms can be uniquely extended to a linear transformation on R^n. Now G/A acts as a compact group of automorphisms on R^n and leaves $\phi(N)$ invariant. By Lemmas 7.2. and 7.3. we find a subspace S invariant under G/A such that R^n is the direct sum of $\phi(N)$ and S. Let E in NF be such that $\phi(E) = \phi(NF) \cap S$. Then $\phi(N) + \phi(S) = \phi(N) + \phi(F)$ and $\phi(S)$ is discrete since $\phi(S)$ intersects the component $\phi(N)$ of $\phi(NF)$ trivially. Obviously $\phi(S)$ is invariant under the action of G/A. Hence S is normal in G. The vector group NE/E has in G/E a compact factor group; therefore it splits as a normal semidirect factor with a compact complement V/E where V is a subgroup of G. Then $G = NV$, and since $NE \cap V = E$, we have $N \cap V = \{1\}$.

The Radical

PROPOSITION 6. *Let C be a compact normal subgroup of G such that G/C is a vector group. Suppose that C is semisimple. Then G is the direct product of C and a vector group.*

PROOF. Let R be the radical of G. Then $R \cap C$ is totally disconnected since C_o does not contain connected normal solvable subgroups. Now $R/(R \cap C)$ is isomorphic to RC/C which is a connected subgroup of G/C and is therefore a vector group. The commutator group R' of R is connected and contained in $R \cap C$, hence it is trivial, and R is abelian. But since it does not contain nontrivial compact connected subgroups it is a vector group. Thus RC is a direct product. Since G/RC is a homomorphic image of G/C on one hand and of G/R on the other it is an abelian connected group and at the same time its component is a semisimple group. Thus G/RC is trivial and $G = RC$.

Let $G \to F(G)$ be a functor assigning to every topological group of some *category* a subgroup $F(G)$ of G. Then $F(G)$ is said to be *fully characteristic*, if for every homomorphism f defined on G we have $f(F(G)) = F(f(G))$.

LEMMA 7. *The commutator group G' of a topological group G is fully charac-*

teristic. *In the category of all locally compact groups the component G_o of G is fully characteristic. Hence, for a locally compact group G and some closed normal subgroup K we have $(G'K/K)_o = G_o'K/K$.*

PROOF. The first statement is well known for the category of discrete groups whence it follows immediately for topological groups. Now let f be a homomorphism of a locally compact group G on a locally compact group H. Every open subgroup of G is mapped by f on an open subgroup of H and all open subgroups of H are so obtained. But in locally compact groups G_o and H_o are the intersections of all open subgroups of G and H respectively which proves the rest.

LEMMA 8. *Let K be a normal compact subgroup of a locally compact group G such that $(G/K)_o$ is solvable and nontrivial. Then the radical of G is not trivial.*

PROOF. If K is not semisimple, then K contains a connected abelian central characteristic subgroup and the lemma is trivial. Suppose from now on that K is semisimple. Assume that among all locally compact extensions of K by solvable groups, G is a counter-example to the lemma such that the length of $(G/K)_o = G_oK/K$ is minimal. Let first $G'K/K = (G/K)'$ be trivial. This implies that G/K is the direct product of a vector group H/K and a compact abelian group C/K; but then, by Proposition 6, H is the direct product of a vector group N and K, and because of the compactness of C, the vector group can be chosen normal in G (cf. Corollary X.3.). Hence N must be trivial since G has no radical. Thus $G/K = C/K$ and G is compact. But then $(G/K)_o$ is necessarily trivial if G has a trivial radical. Therefore $G'K/K$ cannot be trivial. Then $G'K$ is a locally compact connected extension of K such that the length of $(G'K/K)_o = G_o'K/K$ is smaller than the length of G_oK/K. Hence $G_o'K$ contains a nontrivial maximal connected solvable subgroup which is characteristic and hence normal in G_o. Thus the radical of G is not trivial. This contradiction proves the lemma.

PROPOSITION 9. *Let G be a locally compact group such that the radical of G is trivial. If K is a compact normal subgroup of G such that G/K is a Lie group, then G/K is semisimple.*

The proof follows readily from Lemma 8.

LEMMA 10. *Suppose that G is a locally compact group with a vector subgroup V such that the quotient space G/V is compact. Then G/R is compact where R is the radical of G, i.e., G is a C-group (cf. 12.11.).*

PROOF. Let H be a normal subgroup containing R such that H/R is compact and $(G/H)_o$ is a semisimple connected Lie group without compact components. It is sufficient to prove that G/H is compact. The abelian connected Lie group VH/H splits into a direct product of a vector group W/H \and a toral group T/H. Let F/H be a free group in W/H with the same rank as W/H. Then $(G/H)/(F/K)$ is a compact quotient space since $(G/H)/(W/H)$ is compact. By a theorem of Borel[1] the normalizer of F/H in $(G/H)_o$ must be discrete. But this normalizer contains W/H. Therefore $W=H$ which makes G/H compact.

LEMMA 11. THE THEOREM. *Let G be a topological group and V a vector subgroup such that G/V is compact. Then there is a vector subgroup N normal in G such that G/N is compact.*

PROOF. The group VR is solvable if R denotes the radical of G. By Lemma 4 there is a vector subgroup W normal in VR such that VR/W is compact abelian. Then G/W is compact so that W plays the same role as V. By change of notation, if necessary, we now assume that VR/V is compact abelian. Then $R'\subset(VR)'\subset V$, and since R' is connected, it is a vector group. Let now Z be the identity component of the centralizer of R' in G. Since R' is characteristic in R and R is so in G_o then R' is normal in G; therefore the centralizer of R' and its component are normal in G. Clearly $V\subset Z$; therefore G/Z is compact. Let S be the radical of Z. It is sufficient to show that S contains a vector subgroup U such that U is normal in S and S/U is compact. Then, being a compact connected solvable group, S/U is abelian, and by Lemma 4 we find the required vector group N in S normal in G.

Obviously $R'\subset S$ and S/R' is a connected abelian group. Hence every subgroup of S containing R' is normal in S. The intersection $V\cap S$ contains R' and S/R' is the product of a vector group H/R' and some compact group. The vector group can be chosen so as to contain $(V\cap S)/R'$. Now $S/(V\cap S)$ is isomorphic to SV/V because S is σ-compact as a connected locally compact group. But SV/V is compact, whence $S/(V\cap S)$ and $H/(V\cap S)$ are compact. But now the hypotheses of Proposition 5 are satisfied. Consequently R' splits in H; but R' is in the center of Z. Therefore H is a vector group. Since Z/S is compact by Lemma 10, Z/H is compact. This finishes the proof.

REMARK. The statement of Lemma 11 is indeed equivalent to the theorem stated in the introduction if we take into account Iwasawa's theorem (Corollary VII.1).

Examples

EXAMPLE 1. Let T be the rotation group of the plane vector group R^2 and multiply the elements of $R^2 \times T$ according to $(x,r)(y,s) = (x^s y, rs)$. Denote this semidirect product of R^2 with T by $R^2 \otimes T$. Multiply $R^2 \otimes T$ directly with the one-dimensional vector group R. This group G is a semidirect product of the three dimensional vector group $R^2 \times 1 \times R$ by the circle group. Let $f: R \to T$ be any homomorphism. Then the elements $((x, f(y)), y)$ form a normal subgroup H of G without compact subgroups such that G/H is the circle group. Now G is also the semidirect product of the metabelian group H and a circle group.

Thus not every subgroup of a semidirect product of a vector group with a compact group–not even a normal one with compact factor group–need be of the same type. Also, there may be subgroups other than vector subgroups which may figure as a normal factor in a semidirect splitting with the same compact complement.

EXAMPLE 2. Let S be a solenoid and $h: S \to T$ any homomorphism. Multiply the elements of $R^2 \times S$ according to $(x,r)(y,s) = (x^{h(s)} y, rs)$. In the group $G = R^2 \otimes S$ the solenoid acts on R^2 like T and T cannot be split off S. This illustrates the statement at the end of the third paragraph of the introduction.

TULANE UNIVERSITY AND
UNIVERSITÄT TÜBINGEN

A Guide for Further Reading

Since this Memoir was written, ten years have elapsed. During
this period of time we have experienced a remarkable activity in
the structure theory and harmonic analysis of special classes of
locally compact groups with which this Memoir was partly concerned.
It would be impractical, if not impossible, to revise it or even
write an appendix in such a way that the reader could be apprized
completely of the state of the theory today. Nevertheless, it may
be of help to him to have a reading list containing some of the
significant contributions to this area made during the past decade.
In this spirit, we provide the following selective bibliography,
not with the intent to be complete but rather with the hope that
it may illustrate the development of the structure theory and
harmonic analysis of special classes of locally compact groups and
to assist the interested reader to find the relevant sources.
We acknowledge with thanks the advice and assistance of
J. Liukkonen and M. Moskowitz.

Karl Heinrich Hofmann and Paul S. Mostert

January, 1972

NEW BIBLIOGRAPHY

<u>Note</u>: This New Bibliography contains none of the entries which have
already been listed in the Bibliography on pages 67 and 68 of this Memoir.

1 Alfsen, E. M. and P. Holm,   A note on compact representations and almost periodicity in topological groups, Math. Scand. $\underline{10}$(1962), 127-137.

2 Borel, A. and J. Tits,   Groupes réductifs, Publ. Math. $\underline{27}$ (1965).

3 Chu, S.,   Compactification and duality of topological groups, Trans. Amer. Math. Soc. $\underline{123}$ (1966), 310-324.

4 Fluch, W.,   Maximal-Fastperiodizität von Gruppen I, II, Math. Scand. $\underline{16}$ (1965), 148-163.

5 Freudenthal, H.,   Ein Zerlegungssatz für im Kleinen kompakte Gruppen, Math. Arch. $\underline{15}$ (1964), 161-165.

6 Goto, M.,   A remark on a theorem of A. Weil. Proc. Maer. Math. Soc. $\underline{20}$ (1969). 163-165.

7 Grosser, S. and M. Moskowitz ,   On central topological groups, Bull. Amer. Math. Soc. $\underline{72}$ (1966), 826-830.

8 Grosser, S. and M. Moskowitz ,   On central topological groups, Trans. Amer. Math. Soc. $\underline{127}$ (1967), 317-340.

9 Grosser, S. and M. Moskowitz ,   Representation theory of central topological groups, Bull. Amer. Math. Soc. $\underline{72}$ (1966), 831-837.

10   Grosser, S. and M. Moskowitz,   Representation theory of central topological groups, Trans. Amer. Math. Soc. $\underline{129}$ (1967), 361-390.

11   Grosser, S., O. Loos, and M. Moskowitz,   Über Automorphismengruppen lokalkompakter Gruppen und Derivationen von Lie-Gruppen, Math. Z. $\underline{114}$ (1970), 321-339.

12   Grosser, S. and M. Moskowitz,   Compactness conditions in topological groups, J.f.d. reine u. angew. Math. $\underline{246}$ (1971), 1-40.

13   Heyer, H.,   Dualität lokalkompakter Gruppen, Lecture Notes in Mathematics 150, Springer-Verlag Berlin-Heidelberg-New York, 1970.

14   Hochschild, G.,   The Structure of Lie Groups, Holden-Day, San Francisco, 1965.

15   Hofmann, K. H.,   Zerfällung topologischer Gruppen, Math. Z. $\underline{84}$ (1964), 16-37.

16   Hofmann, K. H.,   Über das Nilradikal lokal kompakter Gruppen, Math. Z. $\underline{91}$ (1966), 206-215.

17   Kaniuth, E.,   Die Struktur der regulären Darstellung lokal kompakter Gruppen mit invarianter Umgebungsbasis der Eins, Habilitationsschrift Technische Universität München, 1969.

18   Kaniuth, E.,   Zur harmonischen Analyse klassenkompakter Gruppen, Math. Z. $\underline{110}$ (1969), 297-305.

19   Kaniuth, E. and G. Schlichting,   Zur harmonischen Analyse klassenkompakter Gruppen II, Invent, Math. $\underline{10}$ (1970), 332-345.

20 Lee, D. H.,   Supplements for the identity component in locally compact groups, Math. Z. <u>104</u> (1968), 28-49.

21 Lee, D. H.,   The adjoint group of Lie groups, Pac. J. Math. <u>32</u> (1970), 181-186. (see also Review by W. T. van Est M. R. <u>41</u> #8579).

22 Lee, D. H.,   On fixed points of a compact automorphism group, Mich. Math. J. <u>17</u> (1970), 175-178.

23 Lee, D. H.,   Group extensions by totally disconnected groups, Duke Math. J. <u>38</u> (1971), 205-210.

24 Lee, D. H.,   Group extensions and discrete subgroups of Lie groups, to appear.

25 Lee, D. H.,   On the centralizer of a subgroup of a Lie groups, Proc. Amer. Math. Soc. <u>30</u> (1970), 195-198.

26 Lee, D. H.,   On the homomorphisms of locally compact groups, to appear.

27 Lee, D. H. and T. S. Wu,   On the group of automorphisms of a finite dimensional topological group, Mich. Math. J. <u>15</u> (1968), 321-324.

28 Lee, D. H. and T. S. Wu,   On [CA] topological groups, Duke Math. J. <u>37</u> (1970), 515-521.

29 Lee, D. H. and T. S. Wu,   On the existence of compact open normal subgroups of 0-dimensional groups, Proc. Amer. Math. Soc. <u>26</u> (1970), 526-528.

30 Leptin, H.,   Zur harmonischen Analyse klassenkompakter Gruppen, Inv. Math. <u>5</u> (1968), 249-254.

31 Leptin, H. and L. Robertson, Every locally compact MAP-group is unimodular, Proc. Amer. Math. Soc. $\underline{19}$ (1968), 1079-1082.

32 Liukkonen, J. R., Dual spaces of locally compact groups with precompact conjugacy classes, Trans. Amer. Math. Soc., to appear.

33 Liukkonen, J. R. and R. Mosak, Harmonic analysis of $[FC]^-$-groups, to appear.

34 Moore, C. C., Groups with finite dimensional representations, to appear.

35 Mosak, R., Central functions in group algebras, to appear.

36 Mosak, R. D. and M. Moskowitz, Central idempotents in measure algebras, Math. Z. $\underline{122}$ (1971), 217-222.

37 Omori, H., Homomorphic images of Lie groups, J. Math. Soc. Japan $\underline{18}$ (1966), 97-177.

38 Rider, D., Central idempotents in group algebras, Bull. Amer. Math. Soc. $\underline{72}$ (1966), 1000-1002.

39 Robertson, L., A note on the structure of Moore groups, Bull. Amer. Math. Soc. $\underline{75}$ (1967), 594-599.

40 Robertson, L., Quotients and subgroups of groups with compactness conditions, to appear.

41 Robertson, L. and T. W. Wilcox, Splitting in [MAP]-groups, to appear.

42 Schochetman, I., Topology and the duals of certain locally compact groups, Trans. Amer. Math. Soc. $\underline{150}$ (1970), 477-489.

43 Schochetman, I.,   Dimensionality and the duals of certain locally compact groups, Proc. Amer. Math. Soc. 26 (1970), 514-520.

44 Taft, E. J.,   Orthogonal conjugacies in associative and Lie algebras, Trans. Amer. Math. Soc. 113 (1964), 18-29.

45 Thoma, E.,   Über unitäre Darstellungen abzählbarer diskreter Gruppen, Math. Ann. 153 (1964), 111-138.

46 Thoma, E.,   Ein Charakterisierung diskreter Gruppen vom Typ I, Inv. Math. 6 (1968), 190-196.

47 Tits, J.,   Automorphismes à déplacement borné des groupes de Lie, Topology 3 (1964), Suppl. 97-107.

48 Ushakov, V. I.,   Classes of conjugate elements in topological groups, Ukrain. Mat. Z. 14 (1962), 366-371.

49 Ushakov, V. I.,   A certain class of topological groups, Soviet Math. 3 (1962), 682-685.

50 Ushakov, V. I.,   Topological groups which are almost compact, Sibirskii Math. J. 4 (1963), 689-694.

51 Ushakov, V. I.,   Topological FC^--groups, Sibirskii Math. J. 4 (1963), 1162-1174.

52 Wang, S. P.,   The automorphism group of a locally compact group, Duke Math. J. 36 (1969), 277-282.

53 Wang, S. P.,   Compactness properties of topological groups, Trans. Amer. Math. Soc. 154 (1971), 301-314.

54 Wilcox, T. W.,   On the structure of maximally almost periodic groups, Bull. Amer. Math. Soc. 73 (1967), 732-734.

55 Wilcox, T. W.,   On the structure of maximally almost periodic groups, Math. Scand. 23 (1968), 221-232.

56 Wu, T. S.,   Left almost periodicity does not imply right almost periodicity, Bull. Amer. Math. Soc. 72 (1966), 314-316.

57 Wu, T. S.,   On the topology of a dual space, Mich. Math. J. 16 (1969), 265-268.

58 Wu, T. S.,   A certain type of locally compact totally disconnected topological groups, Proc. Amer. Math. Soc. 23 (1969), 613-614.

59 Wu, T. S.,   Locally compact, totally disconnected, solvable groups, Mich. Math. J. 17 (1970), 69-72.

60 Wu, T. S.,   Discrete uniform subgroups of Lie groups, Topology 9 (1970), 137-140.

61 Wu, T. S.,   On [CA] topological groups II, Duke Math. J. 38 (1971), 513-520.

62 Wu, T. S.,   Locally compact topological groups with finitely generated abelian uniform subgroups, to appear.

63 Wu, T. S. and Y. K. Yu,   Compactness properties of topological groups, Mich. Math. J., to appear.